A Book of Hours

DONALD CULROSS PEATTIE LIBRARY
PUBLISHED BY TRINITY UNIVERSITY PRESS

An Almanac for Moderns

A Book of Hours

Cargoes and Harvests

Diversions of the Field

Flowering Earth

A Gathering of Birds:
An Anthology of the Best Ornithological Prose

Green Laurels:
The Lives and Achievements of the Great Naturalists

A Natural History of North American Trees

The Road of a Naturalist

A Book of Hours

DONALD CULROSS PEATTIE

DECORATIONS BY LYND WARD

TRINITY UNIVERSITY PRESS
San Antonio, Texas

Published by Trinity University Press
San Antonio, Texas 78212

Copyright © 2013 by the Estate of Donald Culross Peattie
Copyright © 1937 by Donald Culross Peattie

ISBN 978-1-59534-158-7 (paper)
ISBN 978-1-59534-159-4 (ebook)

Trinity University Press strives to produce its books using methods and
materials in an environmentally sensitive manner. We favor working
with manufacturers that practice sustainable management of all natural
resources, produce paper using recycled stock, and manage forests with the
best possible practices for people, biodiversity, and sustainability. The press
is a member of the Green Press Initiative, a nonprofit program dedicated
to supporting publishers in their efforts to reduce their impacts on
endangered forests, climate change, and forest-dependent communities.

The paper used in this publication meets the minimum requirements of
the American National Standard for Information Sciences—Permanence
of Paper for Printed Library Materials, ANSI 39.48-1992.

Cover design by BookMatters, Berkeley
Cover illustration: milosluz/istockphoto.com

CIP data on file at the Library of Congress.

17   16   15   14   13 | 5   4   3   2   1

THREE *Ante Meridian*

THREE *Ante Meridian*

ON SLEEP'S fringe, there is a tremulous, mirage-like realm, a long narrow kingdom like Egypt's land, with the shape of a scythe and the feel of a sea strand. It is neither the ocean of oblivion, nor the continent of waking. Here the small waves whisper and flash; the half drowned swimmer lies beached, innocently, with his face in the warmth of sand, and the ebb of sleep lapping his tranquil nakedness. He knows that life is given him back, but is not sure yet that he does not regret the sweetness of death. He hears the insistent whistle of some unknown land bird, who praises the reality to which he will not yet open his lids. He is glad of the bird, and the land, but he clings to the marginal sands of this slender realm.

That realm is neither of Ra nor of Dis. And the hour is not the butterfly's nor the moth's. It is crepuscular, but of the two twilights in the rhythm of planetary rotation it is the more mysterious. Should the dreamer fully awake to the dawn twilight, he would find himself still in a watery half-world, where bats and nighthawks flit. And they say that the great cecropia, the mysterious soft giant of all his tribe, alone of them takes wing in the hour before sunrise. But do fireflies dance again, that quench their light after midnight? Do the vesper sparrows, suddenly, after the night silence, speak again into the thin dark? The midges and mosquitoes, do they hold a second dervish rout, and are there swallows to sweep after them now as at dusk?

But the dreamer does not focus his thoughts on the neglected Nature of dawn twilight. Precision is a beam that will pierce through this precious opalescent light. Only to the calling of the bird, the sweet, reiterant whistle without a name, will he open his ears while the waves lap him. One thing at a time, says the soul, one thing, and very slowly.

For this is the moment when the creative power, the indefinable, uncomprehended imagination, works clay with its lover's fingers. It makes its own dreams; it accepts the sea-wrack forms, the shell-money of Poseidon, the spawn of the subconscious, and, itself alert as never in the waking hours, it weighs each coin, hoarding the precious, shoving dross away.

Imagination knows no limits; it has no shames, and is not even civilized; it is not conscious of itself, calls itself by no names. Just so, the dreamer forgets that he is a man, with a name, with a profession, a certain number of years to his life, a known number in the past, an unknown remainder left to him. He is like a child, yet not a child, for in this hour, lying on that shingle that is his bed, a man has none of a child's bird-like extraversion, nor its rapid avian pulse, nor the power to sleep—dead but warm-dead.

In this hour the faculty called poetry (vaguely enough but perhaps truly) performs its services for the day. Poetry is not the whole of artistic creation, even for a worker in words. Information, organization, even the most finished or appealing style, are matters of craft that only the cleared awakened mind can master. To have something to say, one must endure, one must receive the actual impact of experience upon the senses. Field experience, the naturalists would call it. Brushes with other humans, peeps into books, the passing landscape of the world we move in—these are diurnal realities. They may suffice for very competent creation. But they are not the stuff of imagination.

Whether imagination is a useful or even a safe commodity has been questioned from the beginning of time, and may be doubted to the end. There have been unfortunate men who have so loved or so needed it that, to prolong or to regain the lost twilight strand, they have given their days to chloral, or fled to the last desperate jungle sanctuary of madness. But for the sane the narrow kingdom is sweet. One may not stay long there. The calling of the bird begins to trouble, and the outgoing tide sucks the sand from under the feet. You have to choose between the bird and the depths, and the soul in health will always turn to the light.

The bidden consciousness uprises. The part of the mind that analyses and identifies now names the bird. It is the whippoorwill, a creature of the very

6

Three Ante Meridian

hour, a singer who gives two performances in the cycle, the one in the evening well attended, and the other to a wood where there are few auditors. The dawn song lacks full confidence; it is only a salutation, a recognition of the dusk that dies in light instead of darkness. Already, as the wakening man identifies it, the bird has ceased to trust its own voice. With a few soft last cries it ends its orisons.

The room cups night; it holds a little cavern lake of it. In the window's provincial view of sky there is only a faint shallow luminosity—a delusion of day in which it is not necessary to believe. The shepherd astronomers of the Arabian desert, who named Fomalhaut, Al Tair, Al Debaran, Al Gol the demon with its red eye that flares up or dwindles to the glitter of a wink, they called this hour the False Dawn. The sky is no longer precisely black, but an inky violet, in which flashes the single mysterious jewel of Saturn. Late risen, it lies out on the velvet cloth of darkness, a temptation to the thieving light.

There is no star so startling as a morning planet. For it reminds us that other worlds than ours do not keep our hours nor move upon the rounds appointed to us. In the clear sharpened air they hang out in naked space but visibly nearer to us than the twinkling far-off suns. Either the eye in truth beholds, or the imagination supplies, the actual spherical surface of the sister world. We remember that it, like ourselves, is forever falling toward a sun which perpetually rushes on its unknown errand away from her children. Now, its face glittering with light from the sun, light into which we have not yet rolled, the planet of the rings and the nine moons brings back to the waking man the facts of cosmos—the abrupt and terrible meeting of solid and air, the imperious gesture of gravity, the contrast of light and darkness, each forever victorious.

He stretches in the exultant consciousness of his own two worlds, body and not-body. His name returns; his mind accounts for his family, sleeping about. The dual personality of the imaginative shore is fused into one man; his task, his health, his appetites, his clear vision close up like soldiers on fatigue who are trumpeted into marching order.

But, if he is careful, he has not lost the treasure of the receding hour. Out of the sea, out of the much flotsam and the many shells whose colors fade when they dry, he has kept a single pearl-like thought, and his hands go around it and over it, to polish and remember it. When the day comes, and his head is up and his feet shod, he will work a setting for it, at his bench.

Now he rests, more deeply than before, not sleeping, but quieted and content. While he thinks racing, Saturn spins into the tangle of the window vines. So, a leaf on little earth has power to blot it out. When it is gone the window square is nearly empty; the false dawn is drained out of the bowl of sky and the night bird no longer calls. There is a significant hush, long moments of rest in the music of the sphere, that mark the end of a movement, and the silence of all audience attending upon great beginnings.

FOUR *Ante Meridian*

Four *Ante Meridian*

THERE WILL be a last day for each of us on earth when we shall hear the morning birds, when we bridge for the last time the slender but the very deep chasm that divides their affirmation from the solemnity of the voices of night. (And of these latter note the somber names—screech owl and whippoorwill, black-crowned night heron, nighthawk, vesper sparrow and black-billed cuckoo.) On that day, when the gorge of silence is encountered, it will be a brink without further shore.

To the sick, to the old, to the distraught or the

Four Ante Meridian

sheerly wakeful with that dry cold accuracy of vision that comes to the insomniac, the gap between night and day is each morning consciously bridged. An unwary man in that moment falls most easily into discouragement. It is not a good time in which to remember how you saw some one die, or to recall your shames, or doubt sureness of your footing in this affair of living. Though you scramble up on the opposite bank by daybreak, you will not set out stoutly shod in faith.

I know how often I have waited for the first rainy sounding notes of the thrush to drop distinct and cool after the spiritual drought of that hour. But I had not known of any bird voice in this blank cesura, until I heard one in the cool of the wan hour, in a hot prairie city. It was the city of Lincoln and of Lindsay, and I had come there looking for something of the faith of one, and remembering of the other that I used to know him, and that if he had not lost faith I might have gone to his door and taken his hand. I had found nothing yet, and I was not able to sleep amid the city noises. I felt the pull of the chasm and the many things in the dark that dared me to look at them and then assert what I could still believe. For the poison weed of war

A Book of Hours

grows up to the top of the sky, and men are tortured again for their beliefs, and murderers hold women and children in front of them as victim hostages, and where the great dreams were begotten there is hired corruption. There marches upon us the day when no man will dare to say "with malice toward none, with charity for all," except he twist the words.

There was no light in the room, and I thought, there are no birds to wait for; this is no country dawn. But there was a sound, a metallic timbre as if it were struck out on stone in the street. It increased in volume; the echoes bounded among the building walls. By light hammer strokes they roused me slowly up, first to attention, then to the window.

I saw nothing but the glare of the ground lights and the starless purple of the city sky in the hour just before light. But the notes welled from some mysterious source and perpetually escaped upward. I was sure now that after all it was the cry of birds, the united voice of a flock, that incites itself to ritualistic performance. But I was by no means certain of the bird. In a city there cannot be many species, and most improbably any unknown.

Four Ante Meridian

The last bird was aloft, as I wondered. The purple of the sky became violet, the violet faded to a hint of zenith gray. I waited, and because I had the patience I saw at last the descent. The birds were no more than sooty shadows, silent now, skimming on the tilted plane of their wings, cutting out astronomical geometry in the colorless air. One after the other they would hover over the chimney stacks of the buildings across the street, and then, lifting wings until the tips seemed to touch as those of angels are said to do, each bird would drop with an accurate oscillation into the pit of home.

I thought at first that I had discovered a new song of the chimney swift, and a behavior not before observed. But I was mistaken in this. It is the way of swifts, it seems, each hour before dawn to arise in darkness and, describing fantastic evolutions, spiral up and up to salute morning where the radial lines of light, overshooting the mark of the city, attain unthinkable altitudes in upper space.

Though they lead crepuscular lives, and by preference inhabit darkness, the swifts none the less have a particular name of honor for the day. They have business with the dawn. They take one bath

A Book of Hours

of glory and, if you like to think of it that way, dance once before the Lord.

After the swifts will come the full-throated chorus, the direct and rapturous response of life to the age-old miracle of day. But that is another affair. The swifts rouse in the darkness of their chimney pits at an hour internally sensed (since there is no telling time even by the stars, in the homes they have chosen to make with us) and rising up in faith, in the mathematical certainty of their calculation, they seek a light we cannot see, and glory in it while it is pure, in realms—aviators find them at six thousand feet altitude, flying at one hundred miles an hour—in realms above our dust and our understanding.

What I like best about swifts is that though they have left the hollow sycamores where Audubon found them nesting in a wilderness America, to dwell at peace with man, they are not made ridiculous and dirty by contact with him, as pigeons are, and domestic sparrows. They have the gentle, considerate law of the flock, but none of the lust and cruelty of the pack; they possess solidarity without intolerance. From the jungles where they winter—and no man knows precisely where that is—they bring some natural freshness. There is no ostentation in their faith.

Faith? In what is the modern mind to believe now? We have no awe, who can shatter the rocks and unriddle heaven. But there is a great fear all the same. It is a long time since the Second Inaugural Speech, and longer since the Mount of Olives. There is a sound of Domesday in the air, for man, a gathering of last Apocalyptic armies—the poisoned air, the torturer's laugh, the deluded mob, the corrupted children, the raping in the fair valley on the beautiful day, the piggy eyes and the sagging mouth of the megalomaniac dictator croaking from the balcony, the many theories, the pedantic excuses before and after the fact, economic determinism, dialectical materialism, the divine right of Aryans, the armed right of Japanese, the indecent might of spawning. This is the time for the second coming, O Bethlehem village, O New Salem village where Lincoln listened to the cardinal calling "pretty, pretty, pretty" over the grave of Anne, and read Blackwood, and remembered the chained slaves in New Orleans Market. Where find the faith in that coming, how know in ourselves where in ourselves we will find the white hate of hatred,

A Book of Hours

the at last aroused strength that will thrust shut the doors of hell with a slow dogged back?

The weapons of death belong all to the enemy. If we take them up, we have put on their arms, gone over to their colors. We fight among ourselves, like gladiators for the sadistic pleasuring of Nero. There is no good living except by faith that the world that rolls into darkness rolls out of it again, and that, above the sooty pit we live in, light comes to the zenith, and thence descends to earth.

FIVE *Ante Meridian*

FIVE *Ante Meridian*

Its enormous bulk forever heaving out of nether darkness, the earth is turning toward the sun. The leaf, the flower, the thought, the granite back of the continent, the systolic and diastolic oceans, face another day. Light will act upon each one of them, and cannot leave any of them unchanged. At the least it must take youth from all of them save the ocean—from the ocean it takes the clouds, in the perpetual convection of water.

A thousand miles an hour, day flies the Atlantic. It finds the tossing lightship, and picks up the white signal numbers on its gray flanks. It gives back to lonely driftwood, prophetic of land, and to sargassos of red kelp floating lazy and succulent, their existent shapes. It eats out darkness, leaving only the etched lines of the spars and masts of the fishing fleet. Meeting shore, it runs a finger of cold shine down marginal sand. Now the sea marsh grasses, stiff with the brine in their veins, shake light from their swords, spilling it toward the turf of rush and samphire in long runnels from the gutters of the blades. It smiles wanly on the crooked tidal creeks in their ebb, where the clapper rails prod in the black mud with sensitive bills.

Light finds the great port. It smites glory from the boast of the empty towers, striking on stone and steel and glass, that catch the day while the bottom of the street is still in night. The first busses plunge out from their flocking places in the old square with the arch in it. They climb the midtown hill with a groaning of gears that echoes in the stone hall of the avenue; they charge up the pavement that is glazed with the first watery blur, and, still empty, sweep past the closed museums.

For the moment the hive is cleanly. The day is babe innocent. Not a coin has clinked, not an arm-

Five Ante Meridian

pit poured sweat, not a harlot has combed her hair, not a newspaper been read, not a scale been fiddled, not a scheme hatched.

Now here the star called sun is risen, first a red crescent, then an opening eye, then half a sphere, and at the last, quitting the horizon, it gives the illusion of clearing earthly contact with a discernible bound. Light, direct and ruddy, sweeps down the tree boles of the coastal forests. Five minutes later the miracle is repeated inland, preceded forever by the dawn breeze that finds out trembling aspens and cottonwoods and clatters by in them, running in no other trees.

Everywhere there is a secret withdrawal, a folding up and a putting away of nocturnal things. The children of darkness take themselves off to hideout, bed, and home. The moth goes to the under side of the leaf and, spreading serene gray wings, becomes invisible shadow. The nightcrawlers pour away, visceral-colored and cleanly, into the sod, cunningly plugging their burrows behind them with old leafage. Down every little pit of darkness in bark and stone and earth vanish the darkling-beetles, the rove-beetles, the clean wood roaches. The toad immobilizes at the stump base, the color

A Book of Hours

and the lumpy moist texture of the loam itself. In the hollow of old orchards, the dew is a chill gray lake not yet diamonded and rubied.

Out of the pit files the night shift of the miners with empty pails. Women in gray wait for them, with arms still hugging only themselves and their fear.

Westward, the mirror of Ontario picks up the reflection of the zenith streamers, then almond Erie catches the gleam like humor in a long eye, then woman Huron, with its bays like arms flung open, then the pure pendant drop of Michigan, and last Superior, high and deep and ocean-cold. The wakened gulls sweep off the beaches of the fresh seas; the five mirror surfaces, curved liquids with the curvature of earth, come alight, change from shoreless cave lakes to open blue water combed with the long even summer swell.

The planet spins on, plunging Asia into night, giving back to every prairie grass blade the American day. Farther on, in mining towns that lie in canyon darkness, windows are lit where the next shift gulps hot coffee before going down. Above them, at ten thousand feet, the Rockies pick up the fierce astronomical signals of the sun, and flash rose

Five Ante Meridian

and blue relays from their toppling glaciers. We give back soft answers here, under our envelope of moist and dusty atmosphere. In beauty are daily things begun and finished upon earth, not harshly and forever, as it is in heaven.

SIX *Ante Meridian*

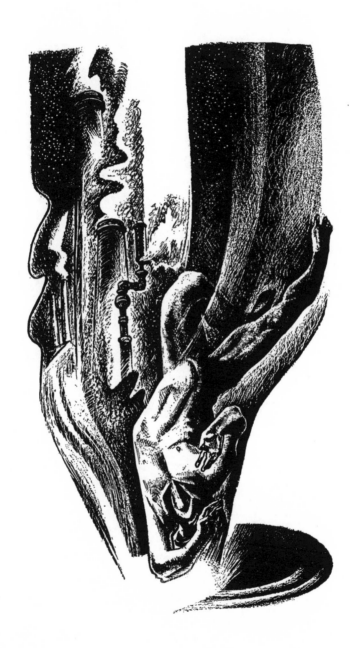

Six *Ante Meridian*

ONE NEVER forgets the sounds that habitually awakened one in childhood. I used to hear the six o'clock whistles blowing off. I knew from where they came—the steel rolling mills. In the noons, in my play beside the shore, I used continually to hear the mile-off mysterious thrumming under the enormous sheds, intermitted by long ringing clashes, as if tons of metal had dropped on a steel floor, and my mind supplied the scene behind the high red fences topped with barbed wire—the small figures running out of the way of inhuman forces, and squealing

A Book of Hours

in all the tongues of Babel. Every night I saw the hell fires from the furnaces, and I came to think of industrial production as a great red flower eternally forced in the black pots of the furnaces, its poisonous petals licking greedily for provender.

Our mills made the steel for railroads in India, railroads for the Tsar, for Dictator Diaz of Mexico. So, said our teacher, they spread civilization into benighted lands that had no steel mills of their own. But there were stories: The Company hid the injured men in its own hospital; it had its doctors sworn to secrecy, coffins all ready, cemetery lots free to their occupants. Men, who in the Company's records had numbers instead of names, fell into molten rivers and became indistinguishable parts of a steel rail, as a bird, caught by a snake, submissively becomes snake, part of its flow and fascination and cold-blooded lust.

You didn't have to believe these stories. But you had to believe in the mill homes that looked builded of grime; you had to believe in the whistles, shrieking for their slaves in the bitter murk of a February morning. I cowered from the open window, nestling under comforters, thrusting my nose into a favorite pillow with balsam needles in it. I was one

of the fortunate, and I knew it then. I deemed myself lucky even in reliably ill health, for it caused
me to be rushed out of winter, and what I woke to
then was the song of the southward wheels.

I remember pushing up the train curtains, and
seeing the pitching mountain landscape, the marching pines, the red clay, the brooks clean and noisy,
the log cabins, and the boys clad only in trousers,
the girl children with skirts to their bare feet, standing in the long early light, waving answer to the
salute of the whistles—good whistles, benevolent
iron dragon that hooted. Good curving rails, blis-
tering with light ahead, with the southern balm.
(It never occurred to me to ask in what mills they
had been rolled.) After that the pines woke me,
and the Carolina wrens, enticing as a brook course,
whistling "WHEE-yoodle, WHEE-yoodle,
WHEE-yoodle!"

I have no nightmare like the horror of becoming
a numbered man. Being screamed awake by a
whistle is integral with this same phobia. Identity
seems to me, and I think it seems to every one not
early flogged out of it, a human sanctity. When a
man arouses from sleep he is instantly himself, and
no other individual in the world. If the convict's

A Book of Hours

number ever really did supplant his name in his waking consciousness, if he ever submissively became part of his costume of shame, part of the bars, then should be well satisfied those who hold that the wicked are not punished enough. Those of us fortunate enough to have committed sin only in our hearts may congratulate ourselves each morning upon the survival of our individuality.

A man arises, and first for a moment is naked—a sensation of pride and freedom that no other animal experiences, since animals have no clothes. True, some cast their skins—insects and frogs and snakes; some molt and shed, but we always discover them embarrassed at those seasons, ashamed and helpless, the birds out of voice, the insects a prey to their enemies. We alone know the Greek glory of nakedness which is not more than passingly erotic.

And in the next minute we assume the human privilege of our personal garments. Women's clothes are more individual than men's, but men rejoice in pockets, and by what a man fills them with before he sets out on the day's adventure shall you know him. The doctor slips his stethoscope into his coat. The thief drops his skeleton key into his

Six Ante Meridian

trousers. The man with the failing heart puts strychnine in his vest. There is the notebook with the cabalistic jottings, the watch with its well-known but never-admitted faults. Knives from cave man days have pointed what emergencies a fellow expects to meet. Variously they betray the clerk with the forest of pencils to be sharpened, the hunter, the angler, the idle whittler, the nobleman, the king's assassin. The naturalist pockets a lens, or binoculars. Last, almost all of us put in something we expect to barter or give away. Tobacco is a social money inherited from savages. A horseman stows sugar about him. You may carry beads for the natives or seeds for the birds. I take bribes for children, that they will go away and let me pretend to work.

Thus deviously equipped, comforted in our ego, we set out to meet fate, and our small personality seems to be built up of such trifles. But, of course, it is not so; it merely flows through them, as life flows through the chemical elements, and employs (rather than obeys) the physical laws. For all creatures have individuality. Any dog-lover, any horse-trainer, any man who ever owned ten apple trees leaning in an old meadow overgrown with Queen-

A Book of Hours

Anne's-lace and ox-eye daisies, knows that individuality is the very essence of living things.

If we cannot discover individuality among the lowlier organisms it may be, in part, due to our lack of interest and our failure to put ourselves in the angle from which traits among, let us say, salamanders, could be thrown into three-dimensional prospective. It would be a rash prophet who would claim that a courteous attention to salamanders would never discover any personal peculiarities among them.

But human separatism has transcended every previous experiment of life in individuality. Identity is the hall-mark of man's genius, one of the goals, we infer, toward which, whether by design or accident, evolution has been tending.

It will be well then to inquire in our neighborhood and discover which leaders and which creeds ask us to lay down our individualities and submerge them in the greater good of the hive, of an ideal state, of a totalitarian state, or a state of grace. An army, of course, submerges individuality, and its excuse is that it could not function if its component parts were allowed to decide whether to stand or run. True—and on armies what a commentary! Dic-

tators require surrender of self, and it is agonizing
to see how easily they obtain it. It takes more effort
to be one's self than it does to be a part of a tramping regiment, or a regiment of women serving only
the womb. Well may any man who loves his self
dread these atavisms—the more, if he loves his fellow man.

Seven *Ante Meridian*

SEVEN *Ante Meridian*

HUMANITY at this hour now divides itself into those who are slug-abeds, and think themselves the finer for it, and those who must arise and shine. Translate the classification into biological terminology and you have the nocturnal people and the diurnal. The former, like some cacti—surly and prickly in the morning—only really flower by night. Gradually with darkening they become human, and by midnight the poets among them are ready to scribble their sonnets; Romeo is now emboldened to swing a leg over Juliet's balcony, and the criminal is at last keyed up to killing Banquo sleeping. The morning temperament is not necessarily better. What are we to think of the man who can murder in the morning? Morning, I've heard, is unerotic and lovers should not meet in it. (No one is obliged to tell what he knows about this matter.) Browning, I believe, wrote poetry in the morning—in the dining-room, as soon as Cook had tidied it.

Mornings suit me best—by preference a fifth-month morning. Never again in the year's round will there be such delicious flexing of the big life-sinews, the long muscle-pull of the instincts demanding to be obeyed, rewarding obedience with pleasure. The immemorial forces—migration, mating, nidulation, and appetite—direct. Protoplasm flows for them, it takes flight, it turns to fight, it seeks to couple, and it devours and assimilates. Infinitely is it bent and adapted and exercised. Of a man this is as true as of a root. Both show the furrows and the knots, the whorls and scars of their imposed experience.

A naturalist could scarce afford to give way to a nocturnal disposition. There are, he knows, two great life peaks in the day-night cycle, when, as it were, Nature changes guard in front of her palaces.

Seven Ante Meridian

One set of organisms relieves the other, and the program is quite fixed; dusk and morning are the scheduled hours. Morning is the ornithologist's great moment, and of a May morning, he may hope to obtain the *optima spolia*. Every bird, migrant and summer resident, is present. A hundred species a morning is a no uncommon achievement in the fifth month for those robustious temperaments bent on the establishing of records. Once past that youthful ambition, a naturalist sets himself to the more significant census of the actual count of individuals. This too will rise to a maximum in this hour and season. Birds on their migrations, weather permitting, most commonly travel for three or four hours at the end of night. Hungry, hunting territorial preserves, eager for mates, they blow in now as the morning mists blow out.

The business of the naturalist is not simply discovery of the obviously new; neither is it merely census-taking; his concern is with the great flow and ebb of the primarily motivating forces. He is absorbed by the complexity of life, and it is not his business, as those who profess the mathematical sciences have sometimes thought, to reduce all phenomena to a few simple explanations. Complex-

A Book of Hours

ity is itself one of the fundamental facts of life; as soon as you try to reduce a May morning to a few chemical laws and physical stresses, you have nothing but those laws and stresses. Life has taken wing. The naturalist knows this just as well as an artist would know it, looking at the same long, slant shafts of vaporous and holy light.

For the origin of the naturalist's science is (though he may have forgotten it) emotional; it rises from appreciation. I am not saying that the naturalist is engaged with esthetics, for they are the science of beauty. The naturalist rather perceives the beauty in science. This has been denied, by scientists who are not themselves naturalists. But the greater the naturalist the less is he abashed by the idea of beauty. Haller and Aristotle, Linnaeus and Humboldt, Darwin and Fabre—they speak with one voice on the matter. Each after his fashion held to the creed of Keats. For Haller, the Alps enameled with flowers were truth, and a great symmetrical and moral cosmology was beauty for the Greek. Darwin found truth and beauty the same thing among the variant primroses beneath English oaks. All these men began in wonder, and arrived at wonder.

Seven Ante Meridian

But day's beginning is peculiarly the property not of the great men but of the common ones. And it is precisely these folk who are most often abroad at this hour. A man, with a day of humdrum before him, goes down for his paper; he opens his front door, and lighter than water but full with the sound and insistence of a sea wave, the May morning washes impetuously in, splashing into the brackish pool of household air with a spray of sunlight and a laughter of fresh smells. He stoops for the paper and straightens up. For a moment he looks back at the day—at the serenity that even street trees have, at the deep perspective of shadow—and shakes his head with a smile half of self-denial, half wonderment. So a sailor, who must keep his mind on the course and bring other souls to land, will have a fond nod for the petrels, who have no destination, no course, no care, no apparent peril.

So there you have it—any morning, every morning, the day I discovered the king rail; the day the prothonotary warbler piped thin gold from the old male willow; the many days when I discovered nothing, because everything was there, complete, miraculous and glistening. The beetles were at toil under the grass roots; the gray striders were in their

45

A Book of Hours

place upon the pond. Hickory bud scales, curling back brilliant and tapered as great petals, released the intricate whorl of the packed leaves for their expansion, for the drawing of their breath.

All things ready, great in their beginning. Nothing in life so beautiful or so promising as beginning. Man's life too looks never so fair as in the outset. On this the pessimist will base his case. He bids you despair because the best beginnings end in darkness spilling before the vision, in the eaten-out shard, the waving triumphant tuft of mold spores. But here he is not talking about life, but death, rather.

For life is anywhere a beginning, as a circle is. Only the human race, with its high anticipations, chooses in the mass to step forth on the treadmill at the brave hour of seven o'clock.

EIGHT *Ante Meridian*

EIGHT *Ante Meridian*

A MAN WHO refuges from the world in some quiet wilderness and finds himself deprived of his morning newspaper will fume, for a few days. He will write to the circulation department, and curse the provincial mails, and stamp his stick in the post office. A man, he cries, must keep abreast of the times—why, it does not occur to him to question. Finally the paper arrives, and somewhere en route Nature has given it a contemptuous battering; the wrapper is torn; the front page is water-soaked. He discovers that most of the news, when he knows that

49

A Book of Hours

it is four days old, has lost every atom of importance.

This lamentable state of affairs continues some
time, before any improvement or cure will be visible. Not that the newspaper ever arrives promptly,
there in Thule; the improvement takes place in the
man himself. He gives up at last, first grudgingly
and finally with indifference. Papers come, and are
opened at leisure, and laid aside for the useful and
noble purpose of lighting the fire. The poor fellow
has now gone completely native; the rains have got
his morale, and like an old house he begins to settle
and accept the elements. His friends, coming to
look him up, on his island, in his mountains, out on
his desert, discover that he is more than a little
touched by his solitary existence. He interrupts
irrelevantly to ask if they realize that they will be
able to see Mercury tonight—it's not usually visible
outside the tropics, you know, except at its apogee.
The current copy of *Auk* is on his table, but he
has not heard that the emperor's mistress is getting
a divorce from her husband. He even asks what
emperor! He has eaten of the lotus. He has forgotten Rome. The emperor takes a new mistress
and in time he learns of it listlessly. There is even

Eight Ante Meridian

a change of emperors, but he can never seem to remember to speak of this new Pompey the Third. It is so much easier to remember Romulus the Tenth; the names of emperors ought not to be varied; underneath the purple they are all the same.

On the road to Rome each morning the busy commuters, as the train glides out of the Sabine Hills, snap open their morning sheets. The sorrowful beauty of the Campagna rolls by the windows out of which nobody looks. For they are all keeping abreast of the times. The Senate is corrupt. Poppaea is going to have a child. A gladiator has written some doggerel that makes the front page. The Teutons are fortifying the Rhine again. The peasants of Dacia are rising against the tithe-gatherers. Drink our Falernian: the best for fifty years. The Britons paint themselves blue with woad. Virgil, noted author, dead.

Poor old Decius, off there on Capri! Writing a book on the conchology of the island, last we heard. Or was it a monograph upon the nightingale? (Buy our nightingales' tongues and our imported Gallic snails; the best for fifty years.) How out of touch with everything he is! Here we are, living in the

most stirring times in history... He who leaves
Rome loses Rome.

But it seems to Decius and to me that of all ways
of beginning the day, a bout with the newspaper is
probably the worst. It is enough to jaundice the
eye for the rest of the waking hours. We see no
validity in the contention that one must acquaint
one's self with the news before setting forth on the
day's adventure of living.

What is the news that will not wait till noon for
your attention? If there is any such, it will be your
private business—your love affair, or a child who
has come to you with a precious project, or a sky
worth dreaming at out the window.

Yes, news by noon is stale. But surely that is
comment enough upon it. In very truth, it would
have been just as stale in the morning. The events
you read of so eagerly happened yesterday, or the
day before, or if it was a Krakatoa that has blown
its head off at the antipodes, last week. No matter;
the editor claps the latest date line on the event
and he hands you the hoax for which you pay him
tithe every day.

I do not belittle the importance or the fascination
of the world's news. The trouble is that there is so

Eight Ante Meridian

little of it—news that would matter as much in Tokyo or Cape Town. Is any other kind well worth remembering, worth learning? For everything we read is a form of learning. We charge our subconscious, already overloaded, with such nasty fare, with the names and the faces of persons whose lives matter to us not at all, and their cock-a-doodle-doings, their divorces and hangings. But, you say, you do not try to remember any of all that. Then why read it, except as a stupifiant? And why take your narcotics in the pearly hours of morning?

Only consider the sands of time! If you spend but fifteen minutes a day with something you do not half respect, you give—let me see—seventy-four and a half hours a year to it. In a life time of reading the newspaper only fifteen minutes each day, you will have devoted to that form of self improvement, would it not be some one hundred and twenty days and nights of your allotted span?

Imagine then that a man, at the final trump, had his one hundred and twenty days and nights given back to him, to use over again as he would. He makes his decision then. He may spend that third of a year of grace over again with the journalists, and

A Book of Hours

we shall allow him to pick the greatest if he likes—Addison, De Quincey, Steele, Greeley or any modern. Or he may dance or wench or drink or climb a peak to see dawn come up out of the plain, or follow an unknown bird or sail around the Cape.

I have no illusions that we would live our lives over better. Sailors would probably put to sea again if they had the chance, drinkers would remain drunk in those precious last months; and, such is the power of stupefiants, newspaper readers would probably while away the tedium of waiting for an inevitable end reading the so-called comic strips without a single smile, just as we see them doing now, all the way from the Sabine Hills to Rome, their eyes traveling without mental break or awakening.

Nine *Ante Meridian*

NINE *Ante Meridian*

THOSE human termitaria, the modern cities, have their own diurnal rhythms. At nine in the morning they have sucked their white-ants, their worker caste of pale-faced, big-headed, small-chested males, and the many females with their immature breasts and pelves, and bodies pallid under their clothes. Only a few belated individuals are seen scurrying to the minute doorways at the bottoms of the skyscrapers. The command that summons them draws away half the adults out of the apartment-cells (which are only the sleeping chambers and nurseries of this

A Book of Hours

social primate). And that command, like the communal will of a true termitarium or ant colony, does not emanate from any central colonial brain or neural plexus. It is not spoken, nor communicated by antennae, since humans have no such useful organs, and it is not even definitely instinctive, which makes it more mysterious even than the directives of other social animals.

So might a philosophic ant write of us, and the complexities of our cities would not baffle her, if we could imagine her with a curious intelligence. At this hour of morning she would impersonally record the flagging tempo of the traffic rhythms, and the long flimsy suburban trains being shunted backward now, empty, out of the termini. The beetle-squat ferries are chugging into the bilgy backwaters of our harbors. In thousands of small shops the spiderweb of necessary or pleasurable or titillating objects is spread, and an alert and hungry merchandiser waits at the center of the wares for the first morning flies. Perhaps the emmet naturalist could comprehend the significance of all this, and of our barracks and our highways, our cattle and house pets, government doles, mendicants, crèches for working mothers, grain fields and grain ele-

Nine Ante Meridian

vators. Such things have analogies in her own civilization; so have wine cellars, royal weddings, airplanes, perversions of the pleasure-rewarded instincts, even revolutions and communism and predatory nationalism. Certainly she could understand work of a physical sort, for her life is nothing else. Tunnels under rivers, cobweb bridges, even prison quarries and chain gangs building mountain roads would not astonish any ant.

This hour is one of the world's hours. It commands; it marches with a great tread, and we seldom escape its cadence. No one but a Trappist has time in it for even a hasty prayer, neither does it proffer contemplation or reveal life's deep perspectives, but is instantly all upon us, a column blaring forth the march tunes of duty, which are bad music but catch the legs and set them to tramping. The very birds are at their business, their beaks too full of nesting straws to sing.

A man who loves his work is a man paid twice over. For the less lucky, work is ransom duly rendered. Done in its season (whether that be the day's midmorning, or the strong years of life), done and well done, it will buy the spirit free for the noon's brief rest, and the long seat upon the rock in the wood when the thrush sings evening. To a man with a great longing on him now for the sea or the beech forest or the moor, these are the hours that turn desire and longing into taut passion and a strong man's will.

He who builds well now, may have something to exchange for freedom. For the idle content of a trout stream in the spruce woods, for the salt hours upon a tilting deck, or a setting forth to bring out of the rainy lost Nantahallas all its fantastic fungus flora. The impulse toward these open pleasures, the right to idle to some purpose—they are instinctive in the boy; we have to promise their satisfaction to him when we propel him to his lessons, reward to come on that sweet future date when books are closed. Even then, once again the young man accepts an extension of the note.

But beware lest we build so high with our quarried stones that we build a prison. And the trapped spirit no more escape—nor even long to.

Ten *Ante Meridian*

TEN *Ante Meridian*

AMID ALL the admirable human industry, the scurry and swarm in the midmorning streets, there are some few who at this dutiful hour appear to dawdle. Many would confidently call them the drones of the human nest. Audubon, gazing out of the window at a vulture becalmed over ragged woods, was idle; Wallace collecting birds and beetles of the East Indies was idle, and idle the theory that broke upon his mind that jungle night when he was afire with sickness. Darwin's work was usually over by ten o'clock. The naturalist seems forever to be committing the high crime of playing on Monday morning.

For, to set forth at this hour—every one else bound inward, he alone bound outward, blithely going against the traffic—looks far too much fun to seem legitimate. The raillery thrown after the naturalist is half envy. Men who pass him laugh at him to comfort themselves, and women bound with children's hands watch him go with a wistful disapproval. We too, they might say to him, would like to make the morning woods our business. To hunt a king rail's nest in the reeds. To pace the beaches, seeking a fortune in shell money, the rose-petal cockles, murex of Tyrian stain, and all the nacre and pearl. We are sorry, they say—irony in their throats—if we appear ignorant because we cannot tell how a grouse makes his drumming sound, or where the moonflower grows. But you must excuse us, for we have work to do. If our leisure began at ten of a fine May morning, we might be disposed, ourselves, to go sounding after lost Atlantis. But, having given hostages to fortune, perpetually we must redeem them.

These friendly stones flung by envy fall short of the veteran and privileged naturalist as the green

Ten Ante Meridian

boughs close behind his back. Others there are who resent him and his calling from some deeper root than a natural jealousy of his good luck. Men who should have known better, like Ruskin and Thoreau, complain against the natural sciences. The naturalist spoils the scene, they say, not only with his vasculum and his field glasses, but still more with his latinity and a confident footfall that treads ground sacred to the poet and philosopher, dispelling the beauty of aboriginal silence. What this fellow knows or seeks to know, partakes for them of sacrilege.

Then there are those less sentimental, more austere, who deny him even the rating of scientist. It was Kant, I think, who claimed that in so far as anything is reducible to a mathematical sentence it is scientific. No more absolutist assertion was ever made by theologian.

Warned off the premises by the esthete who believes that beauty is his business alone, by the mathematician who claims a lien upon fundamental apprehensions, and by the industrious who feel that the world belongs only to the useful, the naturalist must set out upon his morning trail in a humble frame of mind. Men of the sterner sciences, the chemist and the bacteriologist, for instance, could show in an accounting their worth in pence and pounds. But the naturalist's benefactions to mankind are only modestly practical.

In truth, he quietly scorns to be practical more than incidentally. Without pride or shame in this, he classes himself quite outside the servants of mankind's body and economy. He is one with the astronomers and the explorers, the men who, for richer for poorer, have had to go and see.

Men like these know cravings that must be satisfied, commands that they must strive to answer. To find out what is back of the dark swarm of matter in Sagittarius, or beyond the ice mountains, or underneath the stone—this, or this, is their need. Their wonder is the high watershed whence the great streams of religion, art, and science all sprang out of the rock. And deviously rushing down the mountain, each current has sought out a shining but a lonely sea, so disparate from the others that the fishermen in each will say to each other: "Lo! there is no other great sea than this, and no other where one may catch good red herring."

Curiosity, even idle curiosity, is a fleck of divinity in the eye of the beast. When it is not idle, when

Ten Ante Meridian

the strength of an arduous life is in it, it is of itself
a great reverence. And the most scientific virtue in
the naturalist's curiosity is that it is not a reduction
to mathematical terms. It sees true because, like an
ant's eye, it multiplies the images.

For the rest, it has its ethics, but is not canonical.
It serves but has no price.

ELEVEN *Ante Meridian*

ELEVEN *Ante Meridian*

It must be a fine thing to be a Humboldt, and rich and titled, with all life for a productive leisure, to see the glories of the ends of earth. Hooker saw lonely Kerguelen in the antarctic sea, Nepal that looks down on the world from the Himalayan gable. From the sea, through the jungle, to the Andes, traveled Bates in green-waving Brazil. Wallace, old Sinbad of the Indies, had knocked at every island gate, and could speak and write of whole faunas, and postulate extinct continents and land bridges fallen like time's causeway.

A Book of Hours

Such men make great names, in the natural sciences. They have a specimen of any rare thing in their cabinets, and have knowledge of much that is bizarre and all that is significant upon the grand scale. But it is odd how they do not engrave themselves in our affections as do the men who by choice or by straitened circumstance have kept themselves at home. Gilbert White, watching the swifts above Selborne chimneys, Fabre in his weed patch among the shrilling insects of Provence, or Darwin (for he only traveled once, you know) digging up thick Kentish loam to turn the earthworms out and instate them as citizens more necessary to life's economy than man—these pictures of single-handed parochials endure when we cannot remember just where the La Perousse expedition went or what it accomplished or how it came to grief.

We praise these stay-at-homes for the marvels they worked with a biota—a flora and fauna—essentially tepid. Compelled to isolation, deprived of elaborate equipment, away from great libraries, and half cut off even from publication, they cheerfully made the best, we think, of a bad situation. But it is likely that all their obstacles were half a blessing. The blind Huber among his ants, the melancholy

Eleven Ante Meridian

Haller in the Alpine meadow, the lonely Dr. Garden in Charleston who had to write again and again before Linnaeus answered his letters—adversity drove them on, mothering invention and purifying ambition. And after all, to the fresh vision, everything is strange, and may be vitally significant. The great principles of life are all embraced in an acre of commonest earth, no less than in a tropic empire, and it may be, indeed, that they are discovered more truly in the small compass than the great.

Wherever you have been placed, you are already in the field, before you material rich in proportion as your vision is clear. Under the prosperous and platitudinous Illinois farmland lie the great tree fern fossils in the coal mines, all the story of evolution pictured on the black windows of pre-history. On such a marine fauna as is hourly washed up on Long Island beaches Aristotle was never privileged to gaze. There is in all the world no such divine event, so various and so punctual, as bird migration up the Mississippi Valley. And behind the heat and shimmer of a Georgia day, back of the intense blue-green leafage and the shrill of the field insects, are the great features of life, to which for some reason

73

we are purblind—we but feel for them in the sunlight with curious fingers.

Outside any door is the open field. And any season is the time to enter it. It is morning still; it is
the great hour, and light from our sun is raining on
the planet, sifting through the grass blades, absorbed by darkly opaque and uterine earth and con-
verted into unique vital energy that, alone among
mechanisms, grows by what it expends.

Take the least of things; take a blue-grass blade.
Its cells tense with the waters and pressure of living, it captures light, thrusts into the rushing flume
of radiation a silent mill wheel. So a leaf obeys a
star, a star subserves a leaf. Out of inert salts and
solutions the speechless blade upbuilds the tissue
and the sinews of vital event.

In any least field, some vacant lot that would
have sufficed to Darwin or to Fabre, there is this
hardy, irrepressible covering of weeds, each extant
by a chain of miracle, all too travel-ragged to be
lovely as garden flowers are lovely, none without
significance. Thrust in a spade and cut yourself a
life profile through this unkempt sod. How each
weed lays hold upon the old parent earth, suckling
for its specific need! The spade has tumbled out a

Eleven Ante Meridian

white grub of a beetle, root-cutting, assuming that grass was created to feed his hunger. And suddenly, where the spade has torn open a sand pocket, a thin furious skin of ants pours out, an amber species with perpetually subterranean habits, panicked now by the opening of their heavens.

Here is the breath of old loam, the odor of the soil bacteria, enormous for the unicellular, potent, and essential. Here are the earthworm's burrow, the chipmunk's, the streets of the field mice. There's mushroom spawn, laced arabesque design of a geometry wholly irregular. For life is nothing predictable; it is itself the first astonishment, the unlikely thing that actually occurred.

Now in this handful of earth are all life's primal characteristics: the protean shape always individual, the lingering smell of it in the nostrils, the feel of it, resilient, elastic, mucous and colloidal in the fingers. And the way of it, all irrepressible as it is, frail enough, as heaven knows, and no man knows how immortal.

Meridian

MERIDIAN

Noon stands over head, and from frontier to frontier the whistles salute it. So the cock is supposed to blow his horn for midnight. But I know of no other animal than man who makes anything of noon or does it homage. We blast the zenith air; suddenly, in the cities, the offices empty, and out of them the noon-flies swarm. White faces by the indistinguishable thousand show themselves to the sun's yellow face.

Looking down upon the human nest, a naturalist on Venus might suppose that its members re-

sponded simultaneously to the immense phototro-
pism of high noon—the pull of light. Rather are they
abroad in search of that elusive restaurant where
one may fare better for a few cents less. The bright-
winged ones skim the streets looking for it, the girls
clothed in all their salaries, eager with their inex-
haustible appetite for light sweet food and light
sweet living. The men and boys go hunting for it,
driven by a hunger that is nerve hunger; little they
eat will go to muscle.

And there are other cravings; the public libraries
at noon fill up. When I was a young twelve-dollar-
a-week noon-fly, and no bought cooking could
tempt me, I used sometimes to devour a book instead. I remember the beautiful girl, of the Galician
Jewish type, who came there every noon and wres-
tled with the staggering city directory. Most people, if they cannot find a name in the telephone
book, give up their man as lost. Only the lost themselves sometimes learn that every adult among seven
millions is listed in the enormous volumes, his name
and residence and occupation, if any. The history
of all our ephemerid lives is there, and, if you go
from volume to volume, you may trace the nomadism
of each family, from one of our island wadis to the

Meridian

next, the gradual dispersal of its children, the deceases of the patriarchs, the giving in marriage of the daughters. For whom was this Ruth searching? Father, brother, lover—I never learned, and she never found him, not at least in my city noons.

One o'clock is the executive's hour. There are rich menus in it, and there is leisure. Noon belongs to the people, and the food is such as they can pay for and consume in half an hour. Your noon-fly, presumably, is not of enough importance to need time for more than the endless fight against starvation.

But I wanted to mark the meridian with prayer in some form. I used to see people slipping into the cathedral, and out of envy I followed them. But there was nothing inside except a rose-water twilight. I knelt—I have no hesitancy about kneeling anywhere; it is one of the body's natural attitudes, the way to embrace a standing child or to drink from a spring—but no prayers came.

For a year I had been hiding books on ornithology and plants in my office desk. When spring came I threw up my princely job, at a time when a million returned soldiers were hunting for jobs, and went south. The heaviest part of my baggage

A Book of Hours

was my books, and a hand lens and a pair of field glasses with a scratch in the left eye piece.

As I remember those Blue Ridge weeks they seem to me now to have been all noon. I remember that hour for its sheer sumptuousness, its excess of something good. The setting out and the coming back have faded from my memory. Only the noons are vivid because in them I was, myself, inactive. What I collected then that was green and growing was my thoughts.

The best place for noons is on a high rock, and the best attitude for them is the position that lichens assume. On your back, with your hand flung on your brow, you are in a posture of prayer. On your lips you take the communion of vertical light; it stains, as all wines do. So my hand became brown, and I learned a little and did a world of listening and seeing.

There was a valley bell, a very large and old one, that rang news of dinner—golden chicken, and spoon bread, and buttermilk from the spring-house, and soft hill water in the thick tumblers with the horseshoe mark pressed in their bottoms. The noon train, on the other side of the mountain, whistled salute to the zenith hour and from the farm between the brooks came up the sound of cocks blowing.

That was the hour when bushes tucked shade beneath them, like skirts, and nothing escaped the good tyranny of light. You would have had to be a raccoon with a hollow tree at your command, or a beetle under a rock, to find total shadow at that hour. It was, indeed, the hour of the lizards, who shared the rock with me, of the golden wasp who shared the sweets of my luncheon, fanning the mica dust with the propeller breeze of its wings as it hovered, treading air. It was the hour, too, of the buzzard, who shared space and height with me, performing that feat of birds, a sudden upshooting ascent without a wing-stroke. There is no secret in it, of course; they ride to heaven on the strength of the upward column of air from the sun-baked valley floor. First grow your cambering wings, and you may do the same.

Country noons are prodigal of time and economical of shadow. They invite to the sort of contemplating that is done with the head between the knees, with a good view of a single ant, and in the hearing of a brook.

Contemplation is an art not suddenly to be begun. It takes more time to arrive at it than it does to per-

A Book of Hours

form it like a prayer. I am not propounding a literary whimsy; I am talking about the way, it seems to me, that a man may look into himself by staring into the crystal of the world. This is the only safe introspection; the examining of the conscience by night has a great deal too much of the torchlit dungeon and the rack in it. It makes fanatics, and fanatics make life intolerable. Night is a season for music, for love play, for feasting, drinking, dancing, for that blent life of the senses and imagination. These are the ingredients of romance. And night is to lie with. There is no true light in it by which to judge life's colors.

I dare the cynic to try his thoughts out in the noon sunshine that is without shadow. How hard then it is to lock your door, to make a wax image of your fellow man and stick pins in it or melt it! Such stuff is night work. Under noon it is more likely you who will melt, upon your rock, while in the clearing, in the valley below you, man swings his ax—a sun-twinkle and a padded blow—and man's wife, with the sun on her smooth hair and bare arms, walks out to bring him something: a meal in a pail, a kiss, a bit of news about a sick child or a child coming.

ONE *Post Meridian*

ONE *Post Meridian*

DAY STRIDES one pace beyond the zenith. Light still rules, though the sun is on the downward arc of its apparent track through heaven. The heat intensifies; the plain flowers burn; the buzzards look becalmed as if no wind would ever catch their black ragged sails again. Only insect life is roused to some inhuman pitch; the hunting wasps quarter like hawks along the searing ground; the dragonfly darts in ecstasy across the stifling day-miasmas of the marsh, mating in the shimmer, the mate seized like a prey, with her long whiplike body bent upward and backward

A Book of Hours

and carried in the mighty coursing flight, above the closing water-lilies.

For it is the hour of the siesta of the flowers, the beginning indeed of night. On the stagnant waters the lotus and the fatter cow-lilies shut. In the long mountain valleys, stifled between re-echoed heats, many a wild weed shuts its eye; some shift their leaves, drooping them, turning them edgewise, enduring light as if with pain, and assuredly with danger. Clover and oxalis and compass-plant and prickly lettuce all show adjustments to the fierce downpour. On the mountain meadows, uplifted, held high on the granite hands, the delicate arctic-alpine blooms begin to close protectively—glacier gentians, alpine campion, eyebright less tall than the moist black moss.

Almost two centuries ago, in the northernmost university of Europe, Linnaeus drew up his plan for a Floral Clock. At Upsala it was a plan upon paper, a fantasy played upon the then-mysterious theme of the autonomous movements of petals. At Hammerby, his summer retreat, a garden was actually set out, a garden of the hours, where goatsbeard would greet the three o'clock sun-up of that high latitude (Sixty North) and of those long Swedish

One Post Meridian

summer days, and where night-blooming cactus would breathe open, soft and unseen, with the stroke of midnight bell.

Whimsies in gardening were then the fashion; here was a new form of rococo, and not a head gardener or queen but must have a dial of blossoms. But even Linnaeus was not able to construct an actual living garden that would do what he wished, for his numbers do not all bloom at the same time. Good Kerner von Marilaun of Innsbruck composed a much richer and more accurate Floral Clock, corrected for more intensely inhabited latitudes. But they were all pretty follies, and the time came when science learned to be ashamed of such child's play.

In the second perigee of Linnaeus's reputation (it has twice sunk and been thrice shouldered high), my teachers used to point the finger of ridicule at the Floral Clock, and call it in witness of its inventor's essential frivolity. I know enough now to doubt that any but one of them could have been frivolous so wisely, for these Neo-Schoolmen of the laboratory did not know enough about the Hesperidean flora that laughed to the very walls, the red Georgian walls, of our particular Upsala, to have told you how the native vegetables behave in

89

A Book of Hours

the day-night cycle of our latitudes. I doubt indeed that they knew the Linnaean clock save by its reputation, or would have known where to send me, in the old literature of botany, to look for it.

But here I give it back to them, without change save to translate its old-fangled Latin nomenclature into traditional English names. And so you have it just as Linnaeus drew it up for a latitude widely disparate from ours, and with all its imperfections and its lacks:

LINNAEUS'S FLORAL CLOCK

THREE OF THE MORNING

Goat's-beard opens

FOUR OF THE MORNING

Chicory opens
Ox-tongue opens
Hawkbit (*Leontodon tuberosum*) opens

FIVE OF THE MORNING

Tawny Day-lily opens
Iceland Poppy opens
Common Sowthistle opens
Alpine Hawk's-beard opens
Dandelion opens

SIX OF THE MORNING

Hawkweed (*Hieracium umbellatum*) opens
Cat's-ear opens

One Post Meridian

Madwort opens
Red Hawk's-beard opens
Wall Hawkbit opens
Felon Herb opens
Corn Sowthistle opens
St. Bernard Lily opens
Cape Marigold opens
Hawkbit (*Leontodon hastile*) opens
White Waterlily opens
Lapland Sowthistle opens
Fig-Marigold (*Mesembryanthemum barbatum*) opens

EIGHT OF THE MORNING

Poor-Man's Weatherglass opens
Pink (*Dianthus prolifer*) opens
Auricula Hawkweed opens

NINE OF THE MORNING

Wayside Calendula opens   Dandelion closes
Chondrilla Hawkweed
  opens
Red Sand Spurry opens
Ice Plant opens   Goat's-beard closes

TEN OF THE MORNING

Fig-Marigold (*Mesembry-*   Chicory closes
  anthemum nodiflorum)   Lettuce Flower closes
  opens   Corn Sowthistle closes

ELEVEN OF THE MORNING

               Alpine Hawk's-beard
                 closes
               Common Sowthistie
                 closes

A Book of Hours

Wayside Calendula closes
Lapland Sowthistle closes

ONE OF THE EVENING

Pink (*Dianthus prolifer*)
closes

TWO OF THE EVENING

Fig-Marigold (*Mesembry-anthemum barbatum*)
closes
Red Hawk's-beard closes
Auricula Hawkweed
closes

THREE OF THE EVENING

Red Sand Spurry closes
Ice Plant closes
Hawkbit (*Leontodon has-tile*) closes
Fig-Marigold (*Mesembry-anthemum nodiflorum*)
closes

FOUR OF THE EVENING

Cape Marigold closes
Felon Herb closes
Madwort closes
St. Bernard Lily closes

FIVE OF THE EVENING

Cat's-ear closes
Four-o'clock opens   Hawkweed (*Hieracium unbellatum*) closes
White Waterlily closes

One Post Meridian

Cranesbill (*Geranium
triste*) opens

Iceland Poppy closes

Tawny Day-lily closes

Cactus grandiflorus opens

Night-flowering Catchfly
opens

Cactus grandiflorus closes

One with any training for thinking in terms of groups of plants will note at once the singular fact that so many of Linnaeus's golden numbers are twined with members of the chicory family, of which chicory is one of the exceptions that are blue. The rest—hawkbit and hawkweed, hawk's-beard and cat's-ear, ox-tongue and sowthistle and the many more, have the tousled yellow heads of the dandelion and goat's-beard—a weedy and a pushing lot, where it is not the individual petals that shut,

A Book of Hours

but the whole head or capitulum. One might suppose that this was par excellence the family of go-to-bed-at-noons (the name of one of their number). Not so; it is but the dominant family of north Europe. In the matchless flora of Cape Colony it is the characteristic ice plants and fig-marigolds that so behave; in Mexico the cacti tell the hours, above all by night. In our flora we have the uniquely American *Oenothera* whose diurnal members, the sun-drops, laugh in the sun; in the hour of the fire-fly and the thrush song the nocturnal species, the evening-primroses, part lips, soundlessly slip petals, and from a bud become, in the course of a few measures of a thrush's vespers, a gleaming chalice of night, awaiting the nocturnal moths, holding up to dusk the delicate laden stamens, the stigmata that form a cross.

But of whatever flora you compose a Floral Clock—of the gentians of the central European alps, or the South American *Nyctaginaceae* (note the sound of *Nyx*—of night—in them) you arrive, I think, at the same result. Look back upon the Lin-naean clock and observe how all of morning consists in eclosions, free frank expansion, and how from this hour on there is nothing but closure,

One Post Meridian

the sleep—as it seems to us—of the many species sensitive to light. Not until dusk will there again be expansion, secretive then, mysterious, given not to bright colors but to the suspiration of perfume which the feathered antennae of the moths detect far in the forest.

Two *Post Meridian*

Two *Post Meridian*

WHEN IT is two hours past the meridian on the Atlantic coast, then the clocks stand at one in the central valley, and noon rides above Colorado. The continent at that moment is roughly centered in the day. The sun has crossed half way; from ocean to ocean all North America blazes in sunlight, or, even under clouds, is luminous. It moils with the great diurnal activities of its multiplex life forms.

At this moment dawn is rushing across the Pacific, waking strange birds and men, and already night is stealing out of Asia, throwing the shadows

99

A Book of Hours

of Burnham beeches toward the churchyard at Stoke Poges where Gray wrote his condescending elegy for the lower classes. There it is already the hour of the churchly thrush. Here the American flicker larrups the afternoon with his male boast.

Now, then, we have the North American continent in the celestial spotlight. It has the stage, its rattlesnakes and mocking-birds and prairie dogs uniquely performing, somewhere perhaps a brood of its seventeen-year cicadas emerging with a roar. Now its swallowtails, mandarin exquisites of the insect world, too beautiful for toil, emerge in their splendor. Its many unique families of birds are at their curious and various nidulation; its flowers, appertaining to strange, ancient clans, exhale sex chilly and purely.

Here, I say, is this unlikely continent, this America. Here is this spring returned, this day more than half spent, and our hands filled with ancient sunlight that we do not understand, though the leaves can use it. Here are these forests, these lakes great and small, these wild meadows and this earth. And I think that we tramp and stumble through the high, singing groves, following leaders, leaping stiles without questioning their right to be there, cropping close, and going down with foolish bleating to the shearing and the shambles. We act, but so much of it is acting; we go through motions, but they are the gait of somnambulism. Numb to the realities of the big thrusting forces, not half enjoying or even grateful enough for the tender pleasures, we are as wards or minors, ignorant how we are given trinkets in exchange for riches.

Come to! This is the glory hour. This day, this May afternoon, this continent, this modernity, are the stage of life's optimum, perhaps its high tide-mark. Here are migration and nesting, mating and singing, metamorphosis, infinite adaptation, growth and flowering. The great ages were not past geologic time, they are now. Never before have there been such flowers as are in the world this very day, never such singing voices. Never were the knots and whorls of existence so involuted.

Yet one may get himself born out of the womb, one may swink away his life, crack heads, tumble maids, scratch gold under him for a mighty seat, and die old, feared and unrepentant, without any curiosity or need to understand the physiology of thought, the chemistry of conception, the geography of plants, or the tropisms of animals. Plato thought, but without being aware that he did so with his brain; Roland clave paynim skulls without understanding the mechanism of muscles. St. Bernard walked delicately before the Lord, but as he went he shut his eyes to the snowy Alps and the turquoise lake of Annecy, and closed his ears to the beguiling of the birds. For the really orthodox, First Century, Pauline Christianity has no traffic with the perilous idea that Nature is beautiful and good. On the contrary, to be natural is often to sin, even according to modern theology. So the world wagged on without the perspective into the scene that science gives to those who heed it. You may do what Charlemagne and Joan of Arc did, as well or even better, without the self-doubt and the wide tolerance that are the discipline of science.

But can you call yourself a modern? Are you not a walking anachronism, one who can not or will not pick up the tools that civilization has been so long at fashioning? I do not speak of mechanical tools. For one who does not read or write may fire a gun and cut life short in a wild swan or a Lincoln or the child of an enemy. I mean the knowledges, the penetrant daylight way of thinking, the afternoon maturity of this tempered thing that Galileo left us, that Newton and Darwin and Mendel left us. They call it science, and it has many enemies, and they are stronger in the world in our day than in Bruno's. They are burning the books and they hate truth and tolerance.

THREE *Post Meridian*

THREE *Post Meridian*

Y OU CAN talk and talk to a for-
eigner about the Gettysburg
speech, or the Second Inaugural Address, but you
cannot be sure of making him feel, even when you
read them to him, what we feel. To him these are
simply two speeches, very short, very blunt, lack-
ing in figuration, peroration, appeals to action, or
crushing invective. If it comes to oratory, why,
look you, what did this Lincoln fellow ever say
comparable to Napoleon's address to his soldiers
at the feet of the pyramids? "Frenchmen, forty centuries look down upon you..." There's some-

thing to stir the blood! From Demosthenes to Hitler, from Cicero to Mussolini, European history could be fairly well outlined by its oratory; the French Revolution alone produced more good speeches than America in the one hundred and fifty years of her existence as a nation.

To an American, this seems to be the point. There have been so many incendiary, denunciatory, and exultant words and phrases in Old World history that the mental ears are wearied with them, merely to think of them. The applause, the cheers, the *banzais* and *vivas*, if they could all be concatenated as one vast claque and roar and sent speeding on an electric wave, would frighten the inhabitants of Neptune plowing its track through outer darkness.

It is famous that nobody appreciated the Gettysburg speech. By the time the Second Inaugural was delivered, even the Confederacy, even the scornful British, had perceived that here was a man to be listened to, but what heed was paid to those culminating words, "with malice toward none, with charity for all"? By his political supporters none at all.

Very well, replies a Frenchman, you can boast

Three Post Meridian

two good speeches in your history, short and blunt, but, we will say, good. Is that all? Yes, in a way, that is all, and are not two good, living books, well read, better than a library of calf-bound classics that one is taught to respect, but really does not care to read? However, we have some other very good speeches: "I don't see what good shooting him will do him." "And they may keep their horses, to work their little farms." "If we would guide by the light of reason, we must let our minds be bold." "My name is Charles Lindbergh."

Yes, that sort of thing. A joke too funny and too Christian for laughter. A gesture not so much of pity as of sound sense, in the hour when the speaker might have been drunk with victory. A warning to the reasonable that reason is not spirit. Five words of naive modesty. Yes, we count these as national speeches, and even when we explain the historical or incidental associations surrounding them, we know they will not survive transatlantic travel.

I remember, too, the Frenchman who asked, "But is there in America anything beautiful, in the way that the Temple of Love in the park of the Petit Trianon is beautiful?" Of course I said that there is nothing just like that; indeed, though we build a

A Book of Hours

hundred, they are imitations of the real thing, imported as the Italian opera, unnatural as strawberries in January. He shook his head in pity for America, in wisdom; he, of course, knew the answer already; naturally there was nothing so beautiful as that. There is no need for a Frenchman to travel. Or for an Englishman or an Italian.

You cannot compare New Salem village with the Temple of Love, for one has the beauty of a handworn tool, the other of a dainty ornament. Two different spirits breathe out of the soil in the two places. Perhaps I could make a Frenchman see what we see at New Salem, but I would regard it as uphill work, for in a sense there is nothing there to see, no image to worship, only an invisible oracle. But wait a moment, *mon ami; attends,* I have it. New Salem is our Domrémy.

Yet we shall understand each other better by comparing, not leaders and heroes, but countryman with peasant, fisherman with fisherman, and aristocrat with aristocrat. So I shall take you to Marblehead, and you will look at the silvered town, and into the blue eyes in the old men's faces, and listen to the far-off sound of surf, and think of Brittany, and the Bretons. I shall find you an American

Three Post Meridian

château country in Tidewater Virginia. At Charleston you will remember, perhaps, La Rochelle with its lushly green alleys, its mosquitoes, its rotting wharves and lost causes. I shall take you to Hannibal, and Concord, and New Harmony, and Sauk Center, dreaming under mid-afternoon sunshine, as unselfconscious as napping men and women. You will not find them so different, I think, from those towns of yours—say, Maintenon, beside the slow river with the carp and the waterweeds in it.

For, after all, men are not divided by language. They want only some time and opportunity to find between them the common bonds. The ease with which foreigners have comprehended *Main Street* is due, surely, to the obvious fact that England and France, all nations indeed, are dotted over with thousands of Sauk Centers. When my Frenchman asked me if *Babbitt* was a true picture of American life I agreed that it was; I asked him then if *Madame Bovary* was acknowledged in France as veracious. His grin confessed that I had evened the score.

Very well then, thatched Domrémy, Stratford by the Avon, Assisi full still of the poor and the sinful and the barefoot and the sparrows; and you, Plataea, little town that sent every adult male to

A Book of Hours

the battle of Marathon, and you, New Salem, Illinois; we understand each other. Let Rome fall out with Carthage, Tyre with Alexander, Samarkand with Genghis. We know what good may come out of Bethlehem.

FOUR *Post Meridian*

FOUR *Post Meridian*

IF I EVER write, as I always promise myself to do, the natural history of an American parish—or better still, of one square mile of New World earth—I hope I shall remember to take a realistic view of man's place in Nature, for every birth and death, and every spring plowing, is a part of the history that makes natural history. Wherever man goes, he sets up memorials, small or great, of himself.

It is not possible to take an utterly detached view of man's manifold and repetitious doings. But if man is a part of Nature, as inescapably he is, then he deserves, quite as much as a beech wood or a colony of herons or a prairie dog village or a nest of slave-driver ants, an honest treatment by the naturalist. I would go further, and say a sympathetic treatment. I do not mean sentimental. I am aware that the fellow who looks so picturesque from the mountain, with the reins about his neck, sculptured in the immemorial group of horse and share and husbandman, may be a narrow and unimaginative oaf; he may wench and guzzle, and curse the earth and heaven. Often enough he is grateful for little and oblivious of much, and there is no reason to think that his wife and brats will be better. It is not possible to like every one in this world.

But every taste that comes to the palate need not be honeyed. It is enough, it seems to me, that it *is* a taste and that I have a palate with which to sample life's various flavors. Experience is the great adventure, and you are still alive while you have gusto for the new.

So Ploughman Everyman goes into my natural history, disturber, killer, planter, protector of his chosen beasts and grains that he is. He is curious as a raccoon, proud and ungrateful as a cat, I know, intolerant of rivals as the First Commandment. He

Four Post Meridian

is frail enough for his own horse to kill him at the blow of a hoof, but he has the craft of a crow, and is as continuously fertile as the field mice, and persistent and hoarding as an ant.

Many studies and surveys, transects, and statistics of forest and field come to my desk. But in them all I find the learned authors rigidly exclude all man-made environment. They are deeply preoccupied with the primeval biota, and somewhat contemptuously they refer to the alterations due to human intervention, as secondary communities. Of these they dispose in a paragraph.

I do not say that man usually changes things for the better. But surely it is simple realism to recognize that change them he does. I find that many of my betters treat the face of nature as grammarians treat their native language; they describe, rather, what used to be or what ought to be. So the students of politics, before Machiavelli's honest report, described the State as an ideal.

This land is not perhaps what I might wish, if I could have back its buffalo and its elk, its paroquets and sandhill cranes. But this I know, that our presence—mine and my neighbor's—has profoundly displaced an old order. We simply do not dwell in the

wilderness, and the cold and the terrible virginity of
Nature we may not have back.

But I ask myself what I should have to surrender
if we could go back to the wolf nights and the days
of the deer. There would be no old apple orchards,
bent with a century of bearing, hollow of heart
where the bluebirds nest. No cockcrow in the mild
sunlight, to boast: Peace and plenty and long live
the lusty! Neither would the tall weeds be spattered
with the flecks of milk dripping from the deep ple-
thoric udders. No pears flung in fermented waste,
for the autumn wasps to get drunk in. I should miss
the wheat—a burning rim of gold on top of the soft
slope, bowing to the wind, harsh and bearded when
its spikes are drawn through the palm, running
liquid, raising a dust to heaven when the flailing
juggernauts go through the field, yielding a mouse
harvest when only stubble's left.

I am a man; I like the feel of mastery in the
hands, the power to breed the beasts as we wish,
to work them as the admirable slaves they are. I belong to the intolerant European stock and find it
impossible not to like the eyes of my kind,—that slit
of blue in a child's tan face through which one
looks beyond the red barn and the young green

barley, at the sea over which we came—we the men, you the women.

We are the disturbing people who do not accept fate, nor even the initial facts of creation. What we lacked we make. I am not sure that we are, thereby, innocent citizens of the world at all. As creators, we gestate much monstrosity, much imbecility, and crime no end. But we have made beauty, too, we have made the great ideas. We wrote the Second Inaugural Speech and the Brahms' Fourth and we carved the pediment frieze at Aegina, and set up the columns at Paestum and painted *Christ Healing the Sick*, and in our searching we found the periodic table of the elements, and predicted Neptune before it was seen, and gave anesthetics and sterile surgery to the world. Some of all this is already in ruins; much could be forgotten or destroyed; perhaps a little of it is well nigh too late. But we have been a wonderful and a terrible people, who have pried the locks of Nature, looked in the treasure chests, and helped ourselves.

Now we stand in the bold afternoon of our days on earth. We straddle the oceans with legs of bronze, like the Colossus. Our shadow is long into space, for we dream to the stars. Nature will be

A Book of Hours

forced to remember us, by the stripes we gave her, and the mark of the iron bracelet, by the way we stripped her, by the lives she gave for us and the children she bore to us, and the death we died in her arms, spoiled children, after all, children of her own.

FIVE *Post Meridian*

FIVE *Post Meridian*

I T IS many years now, since the last embarrassed poet slunk out of the churchyard, conscious that the world was fed up with elegies and threnodies and meditations upon mossy tombstones. Once almost obligatory subject matter for romantics, the burying ground is now nearly taboo. But fashions need not deter the strong mind. Looking with a biologist's vision, I never pass a country graveyard by. I have no antiquarian interests, and I do not collect mottoes, sentiments, or genealogical information from tombstones. Decidedly I do not want to visit the graves of my friends.

A Book of Hours

Sleeping strangers are the fellows for me. No way involved with them, I take the great testimony of death to heart, without personal pain to that mortal organ.

I say country burying grounds, because a city cemetery, or a war cemetery, is too overwhelming. The only lessons to be learned from a Gargantuan ossuary like Père Lachaise is the sheer number of dead Frenchmen and Frenchwomen there are in it, and the antique moral that all that is born must die. The lesson of the National Cemetery at Fredericksburg seems to be that generals make mistakes whenever they conceivably could do so. And that in war the brave are shot first.

I have never visited a country cemetery by morning, so far as I remember, and not at night. It is in the hour of lengthening shadows, of the still warmed and sumptuous afternoon when the slant rays say *Benedicte* with a long arm, that I go there for more than my soul's good. For something less than a look at my own mortal future, and more than a backward glance at time. I like them for their biographical perspectives.

There is the old burying ground on the hill above Marblehead harbor, for instance, where the whaling captain lies, and his wife, and her ten dead infants. Or the cemetery in western New York where the New Englanders sleep who thought they had got out west. And the colonial Frenchmen in the Illinois village, brown Mississippi eating its way toward their beds. I remember the Huguenots in a Charleston churchyard, the men and women who had heard Richelieu's cannon at the siege of La Rochelle, who escaped death by a bullet and came to die on this hot and foreign shore—of the virus in a mosquito's bite.

Best of all I recollect the burying ground where the slaves slept by their masters' sides. I had been looking for pinesaps, those wizened plants that, leafless and colorless, push up like fungi with the earth still clinging to them, mysteriously to exhale from their cowled cool petals the knowing odor of carnations. I pushed through the hot pine thicket and stepped suddenly into a clearing. Or it had been a clearing once. First the catbriers had reclaimed it, then the sleeping-beauty brambles, then the pines themselves, spindling up out of the breasts of the mounds. There was a fleck of flowering almond left here and there about the graves, and a green running of vinca, but the stones were tilted or fallen.

A Book of Hours

Not a date was later than Manassas. There were names on the white stones. But on the rough field bowlders of the slaves no signature—only the winking of mica and hornblende in the still sunlight. Overhead in the young forest the wind washed with a sweet solemnity—a song like the chorus of an old spiritual, or a whistle between the teeth that might call a coon dog back to heel.

The naturalist, in his preoccupation with the bones and the instincts, the structure and the dance of life, and all its thrust and need, sometimes achieves a reputation for indifference to human event and personal passions. When this repute is earned, the naturalist should be ashamed of himself. If his emotional temperature is too subnormal for the religion and the poetry of human existence, he ought, as a mere scientist at least, to be quick to the natural history of man.

Even viewing man as part of the mammalian fauna, he and his mate and her babies are perhaps the most notable parts of almost any quadrat that an ecologist might select for study. Man, even illiterate man, is anything but a modest citizen of the planet. Wherever he finds himself, he soon constitutes himself Chief Inhabitant. In many ways he is

126

Five Post Meridian

justified. When he ploughs, the field mice have to scamper. When he sows, the crows profit, the obedient wheat springs up. When he goes a-courting, the foxes have some peace for a while. But when he fathers, then the tall oaks crash for a new room to be added on his lair.

And, at the last, a tree gives up its life to make him a home, and a stone is stood on end for him. The last? No, of course, that is not the last of the story. The uneasy earth mound erodes away, in the end. The boards are punkwood and foxfire. With a slow tug of gravity, and a frost heave, earth claims back even her stone, rubs away the graving on it, tilts it, floors it, and finally scrawls her own idea of an epitaph, in lichen runes.

Six *Post Meridian*

Six *Post Meridian*

Now THERE is emptying of the
city centers, the running away
of blood from a congested plexus of the human
body social, a compensatory ebb for the flow
of the morning. Already the buildings are nearly
void. The sun blazes upon the western window
spaces, so that falsely the tall towers seem illumined.
Actually they are less vacant in the night hours,
when room by room the chars are swabbing them,
than at this moment. This is a melancholy hour in
the trade hubs, with the afternoon's news already
dead on the deserted stands, the ticker tape cut off

A Book of Hours

by Atropos, and the look of a dispersed riot in the streets flecked with the mired or tumbled papers, stained, every day, with the blood of the traffic accident, wreathed in the last violet of motor fumes. The executive has departed; only the stale taste of the loss he swallowed this day remains somehow in the air. The as yet undiscovered little peculation of the clerk, the secret tears of the young woman in the rest room, and the great patience of guile and malice, searching all day for catches in a lawsuit—they are left behind like stale breath, waiting for night to black them out, for a sea wind to sluice the city clean of them.

The last of the commuters' trains are puffing and racketing out into the country. The aisles are littered with the tossed aside newspapers. Even the passengers bound for the last stops have ceased to read. They look out of the windows, into the tenderness of the spring evening. Even railroad ditches are beautiful in May; they fling up a spray of their breath, half sweet and half fetid with the peat odor of decay. The blue flags are flowering there in their wide-eyed and natural way, sucking health out of the muck. Above the clattering song of the home rails the swamp tree frogs shrill insistent upon their cool-blooded emotions. Through the opened train window the piping follows, mile long and mile after mile, follows the ear, pursuing with memories of the other springs, when things were harder and sweeter and a man ached, in the first lush emerald of such an evening, for life to begin. His ear is sharper now, his attention unblurred by desires within himself. So he hears, above the bubbling song of the little *Pseudacris* frogs, the soaring sorrowful wail of *Hyla*, prophesying storm.

At their station stops, the men, coming into their kingdoms, swagger home flicking their rolled papers on their thighs; some carry their coats, out of homage to Nature. They hear a cuckoo stuttering in the first shade of the street trees, and the sultry skirl of the black-throated blue, but they do not profess to identify warblers' songs, or even a cuckoo's—only the lawn vespers of a bouncing robin.

At this hour, nearing or retreating, there is always the clatter of children—the haunting yodels by which they signal each other, the light grinding roar of their skates as, on legs still white with winter, they flee anywhere with the wind of nameless spring unrest at their backs till they scud around wistfully where the sidewalks end and the smell of bouncing bet begins.

For women now the evening duties take up their rhythm. In most homes it is the wives who prepare the meals. But in others there is a servant, one of that great regiment of women without leaders or colors, or a full share of human satisfactions. Children, a primitive and candid people, are not ashamed to love servants, for their little bellies' sakes if for no better. They know at least who it is that gratifies the first of the three great instincts, with all the rituals and authority thereto appertaining. Almost every one can remember at least one wonderful serving woman, and the glamor of one kitchen.

This is the hour of kitchens, where woman makes the human magic of cooking, with the aid of her familiar, fire. Often the kitchen is the only room in the house devoid of bad taste and, better still, uninfected by what we are pleased to call good taste —that bad taste of tomorrow. If there is any foyer or shrine or forum left in family life, it is the kitchen at six o'clock. Then and there are exchanged the adventures of living. For once the big children are willing to listen to the little ones, the

Six Post Meridian

tired and parental have an ear for the difficult and nubile. Later, in the freedom of the evening, needs and desires will scatter and assort them by other units. Now you hear all their voices, in chorus. Voices matter of fact, intimate, like the kitchenware plain and beautiful.

The day closes; the sun will set in the marshes, in the frog song and the tall flags wild and common and so deep-rooted you cannot drag them up. As the light goes, as the lamps shine out, one thing becomes plain. Nothing is more important than that the darkness which falls upon these homes should bring no fear with it. For left in peace humanity might take care of itself.

Yes, we know, those lighted windows say, peace will not save us from the agonies of arthritis, the soul attrition of debt, or the icy gaps left by death. But we will keep our troubles. Nobody ever heard us asking some one else to bear them for us; we do not, like the demagogue, demand others to die for us.

They come and tell us of all the wrongs we are suffering, and promise great betterments. Maybe, maybe. But their hate is for us as we are—slow to anger and glad to toil and difficult to split.

SEVEN *Post Meridian*

Seven *Post Meridian*

T HE SUN descends, earth spinning east away from it, west
horizon rushing up to meet it, giving us the most
acute sense of the actual outpouring of time which
we ever gain. The ticking of a clock measures it
away in the style of a metronome, harshly and of
course artificially. Time does not consist in ticks,
any more than a river consists in drops of water;
like a rushing river, time flows. The sands in an
hourglass flow, but their number is disturbingly
finite. The ancients saw long ago that to watch an
hourglass is depressing; there is too much human

symbolism about it to make it a fit measure of so astronomical a thing as time. And the stealth of the gnomon's shadow on the dial is too slow for our patience.

The actual speed of the earth in rotation is comprehensible and graphic only at this sundown moment, and any one can lay hold upon the seeming abstraction and bring it home to himself simply by thinking of the sun at sunset as motionless, and realizing that the horizon is rushing up to eclipse it.

If there are mountains in your view you will see that they are hurled through space, rolling west to east, under their envelope of atmosphere. If you have a plain for your horizon, or if you live where you look west across the ocean, the rim of the world is your precise chronometer. There is a dramatic moment when the fire ball actually descends upon the fields. Now a man may look the lord of day in the face, stare until he burns green eidolons upon his retina. He sees the spherical shape of his tyrant and creator; he feels not simply the circular disk but the globe of the sun, the celestial geometry. Each instant the power of the terrible eye grows less; now only a hemisphere, the red monster is the ember it will look at some time in futurity, when its

Seven Post Meridian

fires are half exhausted and the earth drifted far out into space, its seas withered to stagnant pools, its polar caps creeping toward the equator, its envelope of gases gone, its last volcano impotent of fire and fertile outpouring.

Now the sun is but a third-of-sphere; it has the look of a burning barn in the next township, or a ship at sea afire, the hold a holocaust, only the charred wisp of a horizontal cloud spar precisely cutting it. Then there is only a baleful rim, a final flinging of long ribbons of light toward the zenith, and at last an instant of time, a measure so abstract that it is no time, when men say the sun has set. The forts, around the world, boom out salute, a blank charge; the flags come down like swifts with muted wings, on any longitude every flag at the identical instant, winging home into the male hands that love them.

Officially, night is at hand. In the tropics the dusk is short; the time of the nocturnal creatures is almost come; already it is the hour of the bat and the firefly. On the Pacific the ships, plowing east, speed into night, at their own velocity plus the thousand miles an hour of terrestrial rotation. The brain of the navigator takes the risks head-on, as the prow the spray. He has his instruments, his course, an abstraction with more certainties in it than any of the concrete and treacherous eventualities of shore. North detains the light in summer. There is always a sunset in the east, if you have eyes for it, a counter-phenomenon very characteristic of many astronomical happenings. The zenith is still alight; the mail planes slant in the beams of it. The peaks of the world catch the last of it, while night fills up the world from below, lakes of it welling in the valleys, overspreading the forests and drowning the deserts where not a light shows. From glacier to glacier the final rays are flashed. The unattainable hanging valleys, blue and terrible with crevasse ice, cup up the ultimate radiance, the steep-pitched meadows, the green of their wild sward of flowers laughing with rose, and the cliffs echoing light like sound.

This is the *Gegenschein* of the Swiss and the geographers, not, I should think, to be set over as English afterglow. This light with the unearthly autonomous chalice radiance in it is an up-glow, a glancing from armor of perpetual pure ice, a reflection from Homeric mirrors of geologic element. So each day-night, for eternity, does planet Earth from its upthrust ranges salute that star the sun, Himalayas, Caucasus, Alps, Andes, and Rockies, each in turn flung through the arc of glory, and sent falling over into Earth's conical shadow.

How fair, from Moon or Mars, would Earth appear! A reasonably good telescope would show the dazzling shield of the seas, the spotless virginity of the polar caps, even the dark belts of the jungles and the northern coniferous forests. And, like a glittering spear-head of beauty, like Venus in the sunset glow to our naked eyes, the *Gegenschein* of the Kien Lung in the Mongolian wastes, the twinkling of Kilimanjaro above the tropical night pool, Long's Peak a candle of glory over Colorado where in dirty windows of the canyon towns the lights in wavering strings wink on.

Up there upon Earth, surely, the Beings must be at peace, reverent of so much beauty, so plentifully provided for with every good thing that all could have their share of joy. Lucky, lucky Earthlings, who have those freshening seas, whose mountains are capped in glory and sculptured with the gentle rain from heaven, whose vegetable life grows so tall, forever breathing out the precious gases that its animals require. There, surely they have made an end of misery, abolished sin, achieved mercy and the brotherhood of Mankind.

Eight *Post Meridian*

EIGHT *Post Meridian*

LAST LIGHT fingers the tops of trees, stands them forth in a bathed glory, their heads in day, their long limbs and low boughs already in upwelling shadow. The birds now, you will have noticed, find the tree tops; they like a dead tree best, and they sit in rows upon the spindling twigs and face the spectacle of sunset like an audience. Rough-winged swallows, martins, grackles, and robins, I find, do this. Not one of them that turns his rump upon the lord of day; tranquilly they take the afterglow upon their breasts. The restless tribe—the swal-

A Book of Hours

lows—wheel off in little erratic companies, describing a few banked spirals of joy, to return and edge along the perch and look again. The robin (a singer we have come to underestimate from a deafening familiarity with his lays) will lift up his beak and sing the sun a confident farewell. He has the gifts of the thrush family in his blood, and it is, it seems to me, the finest family of singers in all the temperate world.

Now we would say that the birds were at their reverences, and we would call the robin's slow-delivered syllables a hymn, were we not so certain that no other animal than ourselves has soul or eyes for beauty. This prejudice, I think, is something we have been taught and are afraid to deny. It is not what simple observation suggests to us. In our innocence we once supposed that birds sang when they were happy, like peasants in a Sicilian vineyard. It is true, no doubt, that the singing bird with his voice proclaims his territorial rights, and commands his mate to him. But the voice of a human lover is employed to the same end. A bird in song is, if you like, a creature actuated by instinctive, reflex and tropic conditions so that almost without

Eight Post Meridian

will he sings because he must. To sing because one must—how pure a motive that were! Should human artists obey no other impulse, would art suffer?

The robin sings the sun down, tropic to light as a moth to candle flame. But there's the wood thrush, now, just beginning his song, and he addresses not the sunset but the forest. His voice floats out of the distant chancel like a religious chorister's. It drifts to us through the naves, without haste in the delivery. The notes, spinning one after another through the green clerestories, reach us without the clashing of an echo, for no vault shuts heaven away.

Hear my LAY? The first musical sentence asks the wood for its attention.

We hear your lay. The second line is a response, with the final syllable lowered and affirmative.

Now the singer delivers a long musical sentence, a testament of faith:

True to you, truly DEAR!

Each note shines forth with the chilled beauty of far-off stars, like Mercury, Venus and Jupiter all in the afterglow, sib to each other, part of one perfect creation, but disparate, spaced at visibly different depths in the deepening violet of the sky.

A Book of Hours

Now the voice rises to a triumphant assertion:
Dear to ME!

And at the last, sustained but descending, perfectly rounding to the period of the cinquaine, the
voice lowered without attempt at climax:
True to—

—you.

There falls a rest, perfectly measured and identical with every other interval, the musical value of
silence understood as well as if the singer were a
conscious composer.

I am under the impression that when he is inter-
rupted in the midst of his melody the thrush does
not forget where he left off, but takes up the syllable following the last utterance. We shall not be
the ones to interrupt him. We have leisure to listen
because we take it as a priority of claim on our attention, to listen until the last note has ceased in
the Gothic twilight, has run, a widening ripple of
sound, unhurried, to the mile-off shore that is the
auditory horizon. Like every religious singer, the
wood thrush simply falls silent, and silent is the
sevenfold amen of the leaves. He takes no bow and
expects no applause. And the audience arises without speech and walks quietly out of the woods.

Eight Post Meridian

The first stars prick through, Venus serenely glittering. A star named for its limpid beauty. But she is no naked Aphrodite. She is forever wrapped in impenetrable clouds, unknown in depth, her seas, it may be, still suspended as vapor over the solid core on which we may not gaze.

Between the orbit of Venus in her stage of evolution that we have passed, and that of Mars, a planet that has perhaps lost its seas and clouds so that we actually gaze on its desertic surfaces, spins Earth, miraculously adapted to life.

That adaptation has been called the fitness of the environment, and the fitness of life to the environment is something so perfect that we are driven to conclude that the two mutual adaptations are in essence viewpoints of the same thing. Call this duality accident or design; it totals the same, so somewhere in the staggering mathematics of cosmos you get the equation of life. Once you have it, the number of possible permutations marches autonomously on. Out of the hells of the sun, the needful and terrible emptiness of the spaces, comes all we know,—the infinite individuality of every leaf in the wood, the holy beauty of the thrush's song, the sensitive ear of the listener.

A Book of Hours

Little earth, voice of bird, mind of man, these were not foreseen or foreseeable at creation as an astronomer thinks of the origin of things. Yet they must have been implied, and were perhaps inevitable, in the first premises of Nature.

NINE *Post Meridian*

Nine *Post Meridian*

THE AFTERGLOW is almost drunk away; it is the hour of Mercury always close to the sun, always setting in the horizon damps and dusts just before it becomes dark enough for us clearly to see a star. Mysterious sister-world, far nearer to us than a brilliant giant like our elder brother Jupiter, little world that we can never quite make out, it seems to me to typify the twilight zone on earth and all the crepuscular biota.

The skimming swallow and the wheeling swift are twilight birds; May-beetles bang against the screen, and the big dusty, dun-colored noctuid moths fight to get at the first house lights and perish there. It is the hour of the midge high aloft and of the toad on dewy ground, his tongue long and cunning, swift as the big moths' that hover over the dame's rocket which only at dusk releases its perfume. But not one of these creatures is so skillfully adaptive as are bats to the strange half-light world that is neither honest day nor frankly rapacious night. Whatever is characteristic of dusk life, either in behavior or in bodily structure, bats will have it and to excess, the adaptation carried out to a fantastic extreme.

First of all his tribe to fly, the red bat launches himself into the midge swarms. He is as plentiful as the nighthawks sweeping, calm and graceful, bird behind bird, through the tranquil earliest fall of twilight, as if migrating or going to some particular home. The red bat eddies and giddies and spews himself up out of vortices and swoops at the blackening surface of the pond like a swallow for a drink. He is an honest bat, who shows very little nasty temper and no particular fear of man or day.

But the red bat is the least remote from daylight life, from humanity and what humans consider the rule of reason, goodness, and decency. A few min-

Nine Post Meridian

utes after his launching, the silver-haired bat is in the air. A shade smaller, he is scarcely to be distinguished on the wing; he too is just within the realm of the familiar, and belongs upon the visible side of the twilight time-zone. But when the little Georgia pipistrelle with its weak, fluttering moth-like flight mounts up in gathering dusk forty or fifty feet into the air to hunt the highest midge swarms, we realize that here are animals about which, in uncaged active life, we shall never know much.

If you go to Florida you may meet with wanderers from the great tropic bat fauna—the leaf-nosed fruit bat of the sinister vampyre family, or the free-tailed bat of the Noctilionid family; if you live in Texas you may have learned the Mexican bat by its frightful smell and its platoons sweeping the darkening sky. But even in the candid Middle West you may find that you will encounter bats which still are beyond us. Who has much to tell about Trousseart's bat, Rhoad's bat, Howell's bat, Rafinesque's bat, or the big-eared bat with its enormous auditory organs like tobacco leaves topping its tiny head? Anything, that is, beyond the hunchback mummies lying in museums?

A Book of Hours

For our bats are, in mere species and sometimes in sheer numbers, more plentiful than any of our other mammals, only rodents excepted. Some have practically never been seen to know them on the wing, since they are collected only in their caves where they sleep away the winter. Perhaps some kinds are confined to just a few caves.

Alone among surviving mammals in the well-settled parts of the country do they migrate, as the deer and elk and buffalo used to migrate. Yet no one knows the direction or the rate of their annual hegira. Surely they cannot, when they wish to arrive at a goal a thousand miles away, travel by the same drunken staggers that they use above the garden this night! Like the vanished bison and the antlered lot, they mate in autumn, it is supposed. But how many have ever witnessed that courtship, more fantastic to think about than even that of swifts who "tread" their womenfolk in midair, falling in couples like leaves toward earth with piercing notes, mayhap of the great amazement many another animal has felt.

They mate in autumn, but the young are only now, or even later, brought forth and carried clinging to the teats while the dam flits across the moon.

Nine Post Meridian

The belief among scientists is that in the case of bats, as in some insects, the male germs of life lie dormant from the autumn, through the deep hibernation, and only when ovulation takes place in spring are the females pregnant. Indeed, much about bats—their fierce little hungers, their mask-like sneering faces, their staggered flight, their confidence in dark and air—is insect-like. They are the most *outré*, the most outre-human, at any rate, of all the mammal kind, and yet of them all they are perhaps the most highly specialized and evolved.

If we had the intricate ears of bats we would be able to hear, I think, a spider walking. If we had their electric velocity and perfect correlation of sensory warning, telegraphy to the muscles, and orientation, we would be able to whirl around and stab a pin right through a moth that was passing behind us at the speed of a well-smacked golf ball. And this by starlight. Or to dash blindfold around a room full of hundreds of objects large and small without touching one of them. Long ago Spallanzani showed that a bat can do this in darkness through its marvelous sense of touch, which forewarns it in its rush, by the rebound of air pressure from even so fine a surface as a piano wire.

159

A Book of Hours

The light is vanishing; veils of darkness drop before the failing gaze; in seeming, the fireflies' light is brighter. The whippoorwill slaps the pool of the air with a few tentative cries; in the north he is uncommon enough to be mysterious and disturbing. From the marshes rises the long trill of the toads at their lethargic embraces. Panting and stealthy as Nature by night may be, its ways and its lusts are comprehensible, and its breath is at least hot and its blood red. The blood of twilight is green and its breath chill upon the cheek as a newt's flanks. The creatures of twilight are not distinguished by the great eyes of the owl and the cat tribe, that let in so much light. Bats can scarcely see much better than moles; the nighthawk does not inspect his dinner of midges; he sweeps them into his great mouth in the net of his sensitive whiskers. Twilight asks much special adaptation of its creatures, but it does not require of them to be able to see, nor encourage them in that whim.

The darkness deepens; the red bat has left the skies, and now the silver-haired, who always chooses the middle way, has gone too. When it is just too dark to make out anything clearly, the large brown bat, with his ten inch wing expanse,

Nine Post Meridian

drops before the blurring vision as if the gathering shadows were breaking off twirling pieces of themselves. Last of all, child of Dis, comes the great hoary bat, fifteen inches from wing tip to tip. Big enough to see, surely, but we diurnal creatures have such weak powers of sight! Nightfall besmudges all our judgments, and loosens the control upon our animal emotions.

For scientific phlegm vanishes before this mounting climax to twilight's haggard story. A flake of hell reels over your shoulder, and the breath of his goblin wings stirs your hair. Now indeed you feel a pricking in your thumbs. For these late bats dwell in the farther zone of twilight that begins just beyond your outstretched hand as, poor clumsy ground mammal, you grope amid the trees for home. It ends somewhere in the sorrowful debouchures of the Styx where the big frogs choke. That time-zone is still the most unexplored jungle in all the riddle of common life.

Ten *Post Meridian*

Ten *Post Meridian*

I F THERE were symbols for the hours of day and night, as the zodiacal months roll under the sign of the crab and the scales and the fishes, then this should be the hour of the moth. (Not for any logical reason, I hope, for the essence of a good symbol is that its origins should be forgotten, its application be fanciful, its usage polished by love-without-reflection.) True, most of the pretty species that summon up in humanity that whimsical and tiresome little elf, the collecting spirit, are day-fliers, and others fly at all hours of the night. But this for me is the moth hour,

A Book of Hours

night not yet profound, the dark airs tentative, suggestive upon the cheek.

Brown and silver, gray and black, moon-white and autumn russet, plumbeous, pruinose, rusty and taupe, color of a mourner's cloak and a rustic's hat, the noctuid moths will come to a lantern in the woods, and to a sugared tree trunk where a mixture of molasses and beer will delight their airy appetites and lead them straight to capture. To cold scientific scrutiny, to the clapping on of names with which so far as they know they were not born and to which they certainly would not answer if called.

I am not by temperament a collector, and there are two thousand kinds of noctuids already described from the United States. So, to be honest, I must confess that I have not enough patience with systematics to struggle long with the precise and stilted pedantry of the fine variations in their wing venation. I do indeed see differences in the subdued patterns—the armorial escutcheons and blazonings and bars dexter and sinister, the little false "eyes" that do not see, the small crescentic windows of translucence through which there is no light to shine. Never will any man understand with his heart, if ever with his brain, why these numberless

166

Ten Post Meridian

species flock to the lights to beat out with their dusty wings the nervous tattoo of their perversion for death.

They are subdued of tone, this our most typical family, matching shadow and bark and moonlight and darkness. Offhand, and in the Victorian period of Natural History, one would explain that bright colors would be of no advantage to nocturnal creatures. Vain reasonableness! It merely happens that most of the noctuids hereabout are somber; they would not be so in the tropics. There, exquisite pinks, tiger yellows, deep sea greens and (rarest of all colors among moths) even softly gleaming metal-blues, lose themselves to any sight in the hot darkness under the Southern Cross. Even here, if you are lucky, you will allure any underwing, and by your torch catch a glimpse of roseate and orange, white and yellow, from the hind-wings that, hidden when the creature is at rest, flash in display when it takes its flight.

Rusty is the long chain of reasoning by which the Darwinian period was led astray about the function of beauty in sexual selection. Equally fruitless it were to inquire, What is beauty? For it is a human notion and not a form of natural selec-

A Book of Hours

tion. You may fall in love with a woman for her singing, or for her tresses. But there seems to be little likelihood that a bird chooses her mate for his voice, or a moth her consort for the iridescence of his scales. It is doubtful if choice is involved. By the pull of a beneficial instinct the sexes of my noctuids find each other in the dark; by the fatal proclivities of a tropism they fall helplessly from the faint terminus of a beam of light into its point of greatest intensity. They go in one case to an experience they neither remember nor understand and, for all we know, cannot even be said to enjoy; in another to an event which they will not survive. Of many indeed it must be said that they do not survive even their matings; moths there are which have no mouth-parts; a few nights they live as exquisites of the one appetite; the union over, and the eggs laid, they die in the dew.

Illogical and fated creatures, tragic only to us! Their night world, their wavering dance of life, are so remote from our own daylight labors, that it is righteous in us even to lament their follies. As to the why of them, need there be an answer? It is a first fallacy to ask the inhuman to show good human cause for its existence or behavior. Every

168

one, before he can think biologically, has to un-
learn the reasoning processes that he began to im-
bibe as a child and has subsequently grooved in the
mind as thought habits. Science itself has been four
thousand years at the task; individually we may be
excused for failure, though not for failure to try.

One tries, in short, to school the mind away from
the notion that anything happens, in the bible of
life, in order that something else may happen. The
Bible of religion is partly responsible for our dif-
ficulties in that; things there repeatedly occur
"that it might be fulfilled which was spoken by
the prophets." One has to swallow the fact that the
manifestations of life are not the best or the most
effective or concise or simple that could be im-
agined. The Mississippi could reach the sea in about
half the time that it takes if it would not meander
so foolishly. And man is that divine and impudent
animal who could, if he wishes, dig it a straight
channel (without any surety that the river would
consent to stay in it). But the fact is that the Mississippi flows where it must flow, under extra-
human conditions, and in the same way flowers in
the night bloom when they must, moths must still
come to them, and apparently, though it falls out

169

A Book of Hours

so ill for them, must love the candle and mate with the flame. And life is that energy that will try anything and, eventually, will have tried everything, even those schemes of things that we call beautiful and reasonable.

ELEVEN *Post Meridian*

ELEVEN *Post Meridian*

I T IS not an easy thing, even for an ardent naturalist, to stay up and abroad all night. In the first place there is the cold, which creeps through even summer warmth, and by three in the morning chills to the marrow.

Then there is dew. Every scientist will tell you that dew is a blessing from heaven. It droppeth as a gentle mercy for the little parched field plants that cannot leave their niches and stations to find water, as the animals do. Licking the flanks of the salamanders, who are forever in danger of death by drought of the skin, dew offers them a filmy hdyrosphere through which to migrate to the breeding pools by night. Dew in the woods of a May night is a lake, atomized but stretching to the moonlit horizon. The tall wet grasses overtop the boots and wetly slap the thighs. When I have tried to steal through the marsh grass to creep down where I can hear the soft scolding back-chat of the night-feeding ducks, dew slaps me familiarly upon the shoulder.

Worst of all is my weakness for sleep. A mere habit, of course. Four times I have seen the dawn without sleep, in order to be the first to look on my children's faces. But Nature at night does not prod us awake with fears or expectancy. One may stand enchanted for half an hour in a familiar clearing transformed by moonlight into a ravishing glade of unearthly blue with tender Corot perspectives, but in that lapse of time nothing necessarily happens. The threading fireflies, the antiphonal chant of one frog marsh responding to another, may be all we learn. Lay this down, if you like, to the wretched visual powers of the human animal, to a lack of training for observing night-ways. Or to the fact that as soon as you move, all Nature hears you and flows darkly away to another quarter of the wood.

Eleven Post Meridian

Instead, then, of all-night sessions, I have tried the expedient of sorties at all hours. My sleeping cabin is on the forest's edge; beyond lie the fields, the pond and the many swamps. I am wakeful anyway almost three hours out of any night, and so I slip out sometimes with my electric torch. By its light I can discover the nightcrawlers pouring along in the dew, their gleaming viscera running liquid through the threadform girdles of their muscles. The beam fascinates the screech owls; it is reflected from the great eyes of the noctuid moths, and picks up the opal rage in the orb of a prowling cat, that can pass faint starlight back through the retina a second time, by means of a reflector. Thus is the radiation of Arcturus made to do duty twice for her. I can allure the moths into my cone of light, discover centipedes stalking with a step like wood ash and millepedes that have crawled out of their toadstools. For a moment, in the circle of my torch, I have transfixed a lumbering woodchuck, and terrified a raccoon into indecision.

But still I know almost nothing of night, as every one knows volumes about the biology of day. Perhaps in all I have had a hundred solo hours of night roaming. Any one who could boast as much time spent in a diving sphere could consider himself an authority on marine life in the depths. A hundred hours of solo flying would make me a licensed air pilot. Not so with night. I have never really got into its bathysphere, and, living a night life, sleeping like an opossum by day, prowled the season round in darkness till I extracted ultimate secrets.

Could this be done? It might, if the human animal could improve his sense of smell. By odor, and the power to perceive it, termites and ants know each other in the dark; pack, flock, colony, can unite by the social smell; there seems little doubt that beavers do as much, perhaps birds too, in the dark tree tops, by the oil of their preen glands reunite the migrating flock. Odor is the signal of sex amid deer and all their kind, among the cat tribe, perhaps among the nocturnal monkeys and the alligators, certainly with the moths and beetles. The slug in the night smells food from afar; so do cats, bears, ants, mosquitoes, roaches, beetles, moths, opossums, armidilloes, weasels, rodents, and foxes. Smells forewarn of the enemy's approach, and by night all tainted airs are probably twice as strong as by day. The swift rise in humidity when the sun is gone fills wood and field with waves of odor, as the clearing

Eleven Post Meridian

of the sky of diurnal magnetic storms permits man's radio signals to come in loud, sharp, and significant. A wave of odor in the darkness speaks to an animal as the voice of the President speaks to us by the fireside, as a code raps out a warning to ships at sea. Each animal is possessed of a receptive instrument tuned to perceive a certain range of odors. The snail and the wolf, the sex-hungry specter-monkey and the weasel-fearing vole learn by flair what they need to know, detect nothing else or ignore it. Man-scent seems to be no exception. There is no way to proclaim your passage to the wild folk so loudly as to walk through the night dews.

And before the science of nyctology becomes a thorough line of inquiry, we shall have to learn to hear sounds which now we do not. Tropical naturalists are familiar with the howler monkey; but we are still literally in the dark about the possibility that the night raids of the buccaneering ant are assisted by sounds we cannot hear.

Our sense of touch is still entirely inadequate for night duty, as compared with that of the rat who finds his way in tight black corners by his whiskers. The very facial pits of the nocturnal copperhead are probably more sensitive to air vibrations than the most delicate surface of our flesh. Of the inadequacy of our eyes I need not speak. Objects invariably look to us farther off by night than they are; we do not judge their size or their speed, if moving, with any accuracy.

But perhaps this is all only a way of saying that we are diurnal creatures whose best adaptation to night is the power to roll a stone before the door of the cave and shut out the saber-tooth tiger. With that stone, with the lights we kindle, by our safe sleep and our common front, we not only push back the fear of night, but we have also with these gestures refused to learn more of it.

There have been more thorough and patient attempts than mine to plumb the murky bottoms of night's biology. But even so no one has ever yet made what a Frenchman called, in titling his book, a *"voyage au bout de la nuit."* Shall we translate *bout* as end? Probably we were better advised to call it bottom. Voyage to the bottom of night! Who can say that he has ever descended into those depths, and seen, as the staring eyes of the lynx can see, night's vague yet intricate and monumental landscapes, inked in by a poetic hand? Who has measured, millimeter by millimeter, those chill sub-

Eleven Post Meridian

lunary things, the growth of phallic fungi and the naked protoplasmic creep of a slime-mold? Probably no one has ever even seen a mating of many of our night moths, yet most of them exist, as imagos or adults, for no other act than conjunction.

Night life is day life exaggerated many fold, the factors in it—smelling and hearing, touching and tasting, yes, and even perhaps seeing—far better developed than by day. The amount of mating, of migrating, of eating, of dying, of spore production, perhaps even of growth and the bearing of young, is probably greater by night than by day; the number of mammals and insects that are nocturnal is distinctly a quorum of their respective subkingdoms. So that nocturnal life is very nearly life supreme, and the ways and the forms of day are subdued and attenuated manifestations adapted to the strange and temporary thing called light.

Night comes rhythmically into our experience. Its dark continent full of strange beasts and flowers comes drifting every twenty-four hours so close to the edge of our own bright one, that we might board it by a leap. Yet still we let it every morning float away unvisited.

Midnight

MIDNIGHT

MAN HAS ever had an eye to abstractions. This is unnatural to him, at any rate unlifelike, for living things generally acknowledge only the concrete. Man alone can deal with nonentities like zero and minus. He is astronomically minded. Among plants and animals midsummer's day is not a date on which anything uniquely important happens, but long ago man marked the summer solstice—Stonehenge was built so that the first beam of sunlight on that day should fall directly on the central altar stone.

A Book of Hours

So he greets noon with whistles, though it is
nothing but the drowsy bottom of the golden bowl
of day. And with twelve ringing strokes upon the
bells he startles midnight, when no animal normally
arouses itself, unless it be the woodcock who, I
have heard, then begins to sing. Turning night to
day, man treats his midnights, at least in cities, like
a carnival. The theaters have just been emptied, and
the crowds are on the streets, unwilling yet to sleep,
intoxicate with spectacle and like children hopeful
of a something more. In the heads of the crowd still
crackles the last laughter of the show, still thunder
the proud artist's ovation and the cataclysmic mock
shelling of the mock but prophetic war.

Now, with linked arms, with a gait set to in-
audible, remembered music of the concert, the
throngs sally on the blazing avenues; they wander
into the cool of the cross-streets, hunting intimacy
in restaurants. Everywhere the restless streams, the
jewels and the wagging canes, the newsboys bark-
ing and threading, the performers emerging, just
ready for their fun, never wearied of being looked
at and pointed out. The waiters stand correct at
their tables, the deep kitchens boil with the cul-
mination of their cycle of activity. The wines are

184

Midnight

dragged from their cellars and sacrificed in libation to the hour. Now blooms pleasure, a night flower without visible roots, drawing the moths by its perfume.

This is the hour that beautiful women love as much as they hate early rising. If some day for our fantastic civilization another pattern of life is substituted, some Sparta where women are never allowed outside the bedroom, or some colorless Martian perfection when they are soldiers and iron-puddlers indistinguishable from men, it will be set down in the histories that carefully record only our follies, how we wrote the names of our favorites, our damsel ephemerids, in lights upon the streets. Perhaps even the lights will be half a memory, of the madcap nights when topless Ilium was spangled with a twinkling kingly waste.

For we have mastered darkness as has no other organism. Many times Nature has attempted the solution of the first great problem of biological light—its production without the accompanying danger of heat. In scattered families this difficulty has been overcome, but the experiment has not been repeated since the astounding triumphs of the deep sea fishes. I can think of no luminous birds, reptiles,

A Book of Hours

or mammals save man—who has externalized his light. And now he carries it into the stone guts of earth; he guides his night planes in with beams of it, and passes through walls and flesh a ray he cannot see. Will he send a signal speeding down the light-years? But there is no likelihood of answer. Alone on the dark ocean of cosmos this city ship plows on, with dancing on board.

Man has evolved with time a seamanly rhythm for his city nights. You sense it when you look down from a tower and see the traffic signals change, and the traffic rush and halt and rush on again like the pulsations of a heart. You get it from the crossed arms of the airport lights, waving for its children, and the winking pin-pricks, red and green, of the fan-shaped railroad yards. But it is there, too, in the hospitals, with the nurses pacing the wards, watching for the signs of life and death. It is in the thrum of the power stations, the subterranean clapping and clashing of the newspaper presses, making ready the morning sheets.

Light and communication, pleasure and safety— these seem perhaps like purely material benefactions, and wealth in them is not of itself conducive to virtue. It is obvious enough that crime can avail itself of mechanical equipments that also serve the blameless. Science is not necessarily moral any more than art is. Responsibility is chargeable only to those who employ them.

But for all that, the modern human hive is somehow a moral spectacle, and a moving one. And never more so than in the night, when underneath the play and the display, there runs the great racial vigilance. So that while we sleep, or grow the secret folded rose of the erotic life, or muse alone by the embers, all of civilization holds itself in readiness, prepared to serve like an immense slave. And like a king you may lie down in the certainty that the guard will be changed in the night. If you need a man enough, civilization will call his name through the darkness and anywhere in the immense kingdom find you this one in a hundred and thirty millions. It will bring you his voice, or send him to you on wings.

Whereby it is no secret that I am not one of those naturalists who suffer from cities, or affect to do so, nor do I find a city unnatural or uninteresting, or a rubbish heap of follies. It has always seemed to me that there is something more than mechanically admirable about a train that arrives on time, a fire department that comes when you call it, a light that leaps into the room at a touch, and a clinic that will fight for the health of a penniless man and mass for him the agencies of mercy, the X-ray, the precious radium, the anesthetics and the surgical skill. For, beyond any pay these services receive, stands out the pride in perfect performance. And above all I admire the noble impersonality of civilization that does not inquire where the recipient stands on religion or politics or race. I call this beauty, and I call it spirit—not some mystical soulfulness that nobody can define, but the spirit of man, that has been a million years a-growing. It has shouldered up the concept of divine goodness from the murky notion of a jealous and whimsical old tyrant who struck with plague in the night, to a faith in ourselves to strike back at plague.

It is plain enough how far we have still to go, how much nobler the features of divinity that we might unveil. But, sculptor-like, we first dream them before we unveil them.

One *Ante Meridian*

ONE *Ante Meridian*

AT BEST man is a lonely soul. He bestrides his ball of earth and sees the constellations rising, immemorial, and friendly if he can name them as they wheel impassively above his head. He can, if he is a little clever, tell time by the stars. He knows that, like flowers, they blossom in their season—by winter Orion, with blue Sirius at his frosty heels, and red Aldebaran before him, preceded by the Pleiad sisters, and in summer the serene blue jewel of Vega, Arcturus a lonely topaz, and low in the south, hugging the tropics, the Scorpion with Antares a ruby in its

A Book of Hours

wicked head. As long as man never guessed the distances or temperatures of the stars, they kept him silent company while he watched the desert flocks and wrote a psalm to Jehu who set that firmament to spinning about his untroubled head.

Then Galileo turned his little lens on Saturn, and by bringing the stars nearer, he thrust them farther off. Tonight the astronomers are looking at a nebula five hundred million light years away. Light travels eleven million miles a minute, six million million miles a year. So you may calculate the distance in miles, of this the most remote single object on which the human eye has ever gazed. But the mile no longer possesses use or meaning. And no earthly experience of space supports the staggered imagination. Not even the mind can travel the vacuous black distance to Altair, just rising at this hour, borne aloft on the wings of the Eagle. But the eye sees and the heart greets that golden prick of light through the curtain of infinity. A twinkle, a vibration so fine and high, it seems pitched to the insect's cry from the dewy wild lawns of the night wood.

This, the obedient mind repeats by rote, is in reality one of the less brilliant of the million times

192

a million stars. Yet it burns more than nine times more brightly than our sun, with the power of some 28,000,000,000,000,000,000,000,000,000, or twenty-eight octillion, candles.

And it is to this that man holds up the brief and flickering taper of his solitary spirit.

It is a mere point of view, whether you consider that the enormous candle power of the heavenly bodies should make man humble and afraid, or whether, perhaps, the small and lonely light burning in the brain of the thinker is worth the whole of insensate cosmos. For, about the nebulae contorted with birth, or the dying-out of the giant red ember stars, even an astronomer knows nothing save what he *thinks;* he sees nothing, save by the light of his one candle. He finds "laws" in the coursing of the spheres, as in the behavior of atoms; because these laws (which are actually only descriptions) are the grammar of his own mental language. Science, like art, is an interpretation of Nature; it is not (and even some scientists do not realize this) Nature itself. Thus, at least, speak the philosophers.

To the common man, a million miles is as good as a billion, since he may never traverse them; solids

are still solids, fire still fire, and ice still cold. He
stands in the night and sees the stars, constellations
of the spring night sinking in the west, the prophetic summer constellations rising in the east. He
knows in his bones, no matter what the metaphy-
sician tells him, that when he dies the light of the
stars will not go out for lack of himself to enter-
tain them in his consciousness.

If there are planets outside the solar system, it is
probable he will never see them. If there is life beneath the clouds of Venus, he will not know it.
No, it seems to him that the play of man will be en-
acted only on these boards, this warped and dusty
and glamorous stage, set with the changing back-
drop of the geologic ages. Haltingly and im-
promptu he reads his lines. He has no audience but
himself.

So he lives, our common man, our every man—
because he can do no otherwise—as if the stars were
after all only configurations of fanciful meaning.
As if he had a great destiny, in time and space. All
life is with him, aboard this curious Ark of earth.
It breathes and runs and flings out its spores in the
night and with feathered antennae senses significant
odors through the dark miles, precisely as if it were

One Ante Meridian

not limited in its rounds to a zone as fine as a circlet of thread. It denies death, by every birth, and by every existence builds as if it were possible for anything to be built to last. Like an aristocracy, the organisms have arrogance to match their frailty; though they came up from the clods, their every gesture proclaims the pride of race, the beauty of form, the rapture of living.

Man turns him to sleep, that he may economize upon the little candle of the mind. He steps inside, and bolts the door against the cold hells of space, the searing paradises of astronomical light. He treads the steps, up to his bed. And stands a while, to watch the quiet breath of his companion. She is there, living and warm, yet remote, too, already slipped beyond him into peace. Sleep; what is sleep? Something more difficult to define than death, a twilight that, some night, will have no further wakening shore. Man draws the burden of companionship between his arms, and clasps it close and certain, before he blows this long day's taper out.

Two *Ante Meridian*

Two *Ante Meridian*

Now is that hour that, in a gloomy view of Deity, has been called God's. For it has been said by generations of old wives, even by nurses and by doctors of an era past, that at this moment the soul most easily quits the body, the body having within it metabolic tides. This is the ebb, then, if we believe with some; this is the cesura, the pause, the invisible suture where the closed circle of life was welded in the fires of creation. Here the chain of the body breaks, at its weakest line of stress, and the spirit steps down into the chill Styx. Or it wakes to perpetual morning, if you prefer. In either event, it quits the known; it comes not back. The weary, the dusty, the tormented escape. But the young and the fair, the precious and the tender, are driven out of their fleshly habitation, evicted from life to mysterious destiny, most ruthless at this nadir hour.

I believe I am right when I assert that hospital statistics do not bear out this supposition. There is no one hour of the twenty-four at which most souls slip away. In these matters, as in others, death is impartial. It recruits by conscription, and its ranks are never filled. The wise, perhaps, are submissive, but, submissive or rebellious, we are propelled by a great hand against our naked shoulder.

The body, of course, cries out: But not my life! The heart that loves cries: But not their lives! I am as vainly passionate here as any other; I find, upon self-searching, that I cannot honestly believe that all lives are equally precious. Who shall be the judge? I do not know, but I say, beware of the man who affects to tell you. Beware of the war-lords. Beware even of him who proclaims that God is the inscrutable judge who, if He slays the young and innocent, and lets the evil flourish, knows best and will act in His time. That is jungle-thinking, and

Two Ante Meridian

every bone in me refuses it, hymn, cant, and voo-
doo. Thank you, I shall fight for the child's breath.
Civilized, I may tolerate the Martian monster sit-
ting astride the tank—though it is an open question
whether civilization dare tolerate him—for I have
not much violence in my blood. I have faith in
science. I believe in the reality of ethics. I have be-
held beauty.

And you believe on these things. You imply your
act of a modern's faith when you call the doctor,
when you accept the certainty of the astronomer's
calculation of the date of an eclipse, when you lie
down to sleep in the assurance that those to whom
you have given your trust are not wolves who will
kill you in the night and put their mouths to your
throat. The whole fabric of human life—with
whose order and pattern no other animal life is
momentarily comparable—is based upon these posi-
tive assumptions. The healthy soul is not passive,
and man is not a fallen angel. But he may just pos-
sibly be on the evolutionary road toward angelic
transmutation.

For he can hear the long trumpets; he lifts his
head to listen. He beholds a few who stand out far
ahead; the best of the rest follow, stumbling, by

A Book of Hours

their lights. Defiance of death is but the most dramatic of heroisms. Every tolerant mind is heroic. All strenuous thinking is heroism; hourly and daily people less fortunate than I live more manfully than I, with less of self-pity and self-conceit. They accumulate the real wealth of the world, not for themselves but for the hive, the golden store of life's goodness.

So man, who comes out of darkness, goes not into it; only the individual body rightly travels on that road. He goes toward the open, the great free steppes—a spirit mounted upon stallion body, a tireless rider who swings from steed to steed.

Donald Culross Peattie (1898–1964) was one of the most influential American nature writers of the twentieth century. Peattie was born in Chicago and grew up in the Smoky Mountains of North Carolina, a region that sparked his interest in the immense wonders of nature. He studied at the University of Chicago and Harvard University. After working for the U.S. Department of Agriculture, he decided to pursue a career as a writer. In 1925 he became a nature columnist for the *Washington Star* and went on to pen more than twenty fiction and nonfiction books over the next five decades. Widely acclaimed and popular in his day, Peattie's work has inspired a modern age of nature writing.